T0135870

Bibliografische Information der Deutschen Nationalbibliothek

Die Deutsche Nationalbibliothek verzeichnet diese Publikation in der
Deutschen Nationalbibliografie; detaillierte bibliografische Daten sind
im Internet über http://dnb.d-nb.de abrufbar.

ISBN 978-3-8325-2025-0

Logos Verlag Berlin GmbH
Comeniushof, Gubener Str. 47,
10243 Berlin
Tel.: +49 030 42 85 10 90
Fax: +49 030 42 85 10 92
INTERNET: http://www.logos-verlag.de

VRIJE UNIVERSITEIT

3D multi-scale finite element analysis of the present-day crustal state of stress and the recent kinematic behaviour of the northern and central Upper Rhine Graben

ACADEMISCH PROEFSCHRIFT

ter verkrijging van de graad Doctor aan
de Vrije Universiteit Amsterdam,
op gezag van de rector magnificus
prof.dr. L.M. Bouter,
in het openbaar te verdedigen
ten overstaan van de promotiecommissie
van de faculteit der Aard- en Levenswetenschappen
op maandag 13 oktober 2008 om 13.45 uur
in de aula van de universiteit,
De Boelelaan 1105

door

Thies Joachim Buchmann

geboren te Hamburg, Duitsland

promotoren: prof.dr. S.A.P.L. Cloetingh
                 prof.dr. F. Wenzel
copromotoren: dr. F. Beekman
                 dr. P.T. Connoly

The research reported in this thesis was carried out at the

Department of Tectonics and Structural
Geology
Faculty of Earth and Life Sciences
Vrije Universiteit
De Boelelaan 1085
1081 HV Amsterdam
the Netherlands

Geophysical Institute,
University of Karlsruhe
Hertzstrasse 16
76187 Karlsruhe
Germany

Financial support was provided by the Geophysical Institute at Karlsruhe, Landesstiftung Baden-Württemberg (project 591), DFG (Collaborative Research Centre/Sonderforschungsbereich 461, project A6) and the Chevron Corp.

CONTENTS

ACKNOWLEDGEMENTS

In the first place, I want to thank my promoter Prof. S.A.P.L. Cloetingh for the opportunity to defend my thesis at the *vrije universiteit* Amsterdam. Furthermore, I want to thank him for his support and for his encouragement to publish and to finish this thesis.

I also want to thank my promoter Prof. Friedemann Wenzel for the opportunity to perform this research in the stress group at the Geophysical Institute of the University of Karlsruhe. Especially, I like to thank him for trust in me to manage the research and administrations on my own and for the fruitful discussions.

In addition, I like to thank my two copromoters for their help and constructive criticism. I would like to thank Dr. Fred Beekmann for his comments and reviews during the final stage of this thesis, which improved the work a lot. Dr Peter Connolly was my "supervisor" both, during my work in the stress group at Karlsruhe and during my internship at Chevron in Houston. I want to thank him for all the effort and enthusiasm he invested in this work. Furthermore I would like to thank him for encouraging me to improve my written and spoken English (the … is freezing cold). I hope we will complete many more projects in the future.

I would like to thank the members of the reading committee, namely Prof. Roy H. Gabrielsen, Prof. Peter Ziegler, Prof. Jan Diederik van Wees, Dr. Matthias Gölke, Dr. Jose Dirkzwager and Prof. Karl Fuchs. Their thorough revisions and constructive comments helped me substantially to improve the manuscript and finish this thesis.

Furthermore, I like to thank the current and former colleagues from the stress group at Karlsruhe for the very pleasant work environment (BBQs, cocktail parties etc,) and the constant opportunities for discussion and collaboration. Namely I want to thank: Junior, Mbach, Verena, Andreas Barth, Daniel, Gwendolyn, Tobias, Birgit, John, Oliver, Paola, Wuestefeld, Philipp, Tingay, Alik, Karl Fuchs and whom I forgot for any reason. I want to thank Jose Dirkzwager for the good collaboration, her constructive criticism and the good times we all had together with Peter and Gwendolyn in Karlsruhe, Houston etc. Last, but not least I thank Andreas Eckert for all the discussions and the interesting questions we worked on. Have fun at Rolla and train them how to drink Weizenbier and BBQ!

I like to thank the Rock Mechanics Team of ETC - Drilling & Completions, Chevron Corp. for the opportunity to do an internship in 2007 with them, during which many of the workflows developed during the work for this thesis could be improved and evaluated using oil field data. I would like to thank the members of the Rock Mechanics Team for the constructive discussion during the work for the individual projects.

In the end, I like to thank my parents, Renate and Joachim Buchmann for their care and support during the last 3 ½ decades.

I want to thank Gwendolyn for all her patience through the time I have spent on this thesis and for her patience with me in general. I hope, in future we will not be divided by irrelevant obstacles such as the Atlantic Ocean or the stormy North Sea and can spend more time together.

SUMMARY

The Upper Rhine Graben (URG) forms the central part of the European Cenozoic Rift System, an intra-plate rift located in the Alpine foreland. The development of the URG is linked to collisional deformation in the Alpine foreland accompanied by significant Moho uplift, with the maximum uplift being located in the southern part of the URG. The URG is currently reactivated as a sinistral shear zone driven by far-field loads (i.e. Alpine collisional and ridge push forces). This leads to relative slow tectonic deformation accompanied by low to medium seismicity but a significant amount of ongoing subsidence of several sub-basins (0.1 to 0.2 mm/a geological, 1 mm/a geodetical rate). The understanding of the present-day kinematics within the URG is still limited and only scattered data sets of stress indicators and in-situ stress measurements are available. The seismicity of the area is widespread and not restricted to the border faults of the graben. Few faults are known to be seismically active. However, for the majority of faults no information is available on their Quaternary activity or on their reactivation potential under present-day stress field conditions. The aim of this thesis is the simulation of the present-day crustal state of stress of the URG area using a multi-scale 3D finite element modelling approach and thus to provide data for a better understanding of the recent kinematic behaviour of the URG and the reactivation potential of the graben faults.

Three dimensional finite element analysis (3D FEA) is a tool that can provide detailed constraints on the kinematic behaviour and estimates of in-situ stress magnitudes and orientations of a geological system. The method of FE modelling is applicable to analyse the tectonic behaviour at various scales ranging from plate-wide models to local scale models. For this thesis, a multi-scale approach is chosen. This approach includes a regional scale model covering the entire URG and shoulder areas. Additionally, the kinematic behaviour of two sub-regions within the graben (central and northern graben segments) is investigated with two sub-models of higher spatial resolution including a large number of fault surfaces. In the framework of this thesis, a method and work process is developed for the construction of these complex model geometries using commercial software packages (gOcad®, HyperMesh®, ABAQUS™). The analysis and interpretation is performed using the in-house post processing package GeoMoVie©.

In order to construct and solve a 3D FE model many different data types such as the geometrical structure, material properties and boundary conditions have to be integrated. The first part of the regional modelling study includes a definition of the elements and geometric structure necessary to reproduce the tectonic behaviour and to quantify the contemporary crustal stress field of the present-day URG. The regional model contains three lithospheric layers representing the upper mantle, the lower crust and the upper crust consisting of shoulder and graben units. The varying thickness of these layers is implemented in order to analyse the effects of gravitational potential energy differences on the present state of stress in the URG area. Lateral depth dependent density variations are assigned to the lower crust and the shoulder and graben units in order to account for the complex Variscan structure of the model area. The lower crust and the lithospheric upper mantle are implemented in order to apply an unevenly distributed vertical load as a function of the Moho topography on the base of the upper crust. This vertical load is generated by gravitational unloading of the upper mantle. The graben is bounded by listric border faults, implemented as frictional contact surfaces.

In order to simulate the crustal state of stress for the URG region, appropriate boundary conditions have to be defined. In this study, displacement boundary conditions are defined at the sides and base of the model. Gravitational acceleration is acting on each element directed towards the Earth's centre. For the simulation of the kinematic behaviour of a tectonic system using a numerical model it is insufficient to consider only the boundary constraints active during the modelled time span. Rather, it is necessary to define an initial state of stress that mimics the natural conditions as closely as possible and then load this "pre-stressed" model with the appropriate kinematic and/or thermal loads. In this study, a spherical loading procedure and a pre-stressing approach is applied in order to account for the elastic compaction and to approximate the in-situ stress state, assuming hydrostatic pore pressure and the presence of residual stresses. The loading is conducted in several steps including elastic compaction of a generic spherical model, gravitational pre-stressing and tectonic pre-stressing of the regional model (global model). The stress state predicted by the regional model is calibrated using available in-situ stress estimations and stress indicator data. For the loading of the local scale (sub-models) the obtained basal and lateral displacements from the global model are applied as boundary conditions. For benchmarking of the results from the sub-models, the vertical surface displacements predicted are compared to the regional uplift pattern, local fault slip data and the distribution of areas indicating increased recent tectonic activity. Furthermore, predicted fault slip rates are compared to available geological and geodetical data.

The results of the regional model study reveal a significant influence of the various topographies (surface, Conrad and Moho), as well as of the geometric structures implemented on the present-day stress state and the recent kinematic behaviour of the URG. The boundary condition analysis of the URG model shows that the amount of assumed Moho uplift is a critical loading condition on the regional kinematic model. The modelling results suggest that the URG is currently being reactivated as a sinistral strike-slip system with the central segment of the URG forming a restraining bend. The overall tectonic regime in the area is transtensional. Strain partitioning is predicted for segments of the border faults.

The results of *sub-model A*, addressing the central segment of the URG, indicate that within the present-day stress field, this segment is more compressional than the remainder of the graben. Highly variable faulting mechanisms including extension, strike-slip and inversion of pre-existing extensional faults are predicted for this area. The comparison of seismogenic faults in the area with the results of a slip tendency analysis suggests that damaging earthquakes in this region are possibly associated with not favourably oriented fault segments.

Sub-model B covers the northern URG segment and consists of several fault-blocks and sub-basins. The results for this model area indicate that within the present-day stress field, the northern URG segment forms a releasing bend. Here, the lateral compression on the graben is reduced kinematically due to a change in strike of the graben. The segment of the Eastern Border Fault, which is predicted to be most active, coincides with the location of the most pronounced Quaternary depocentre. Furthermore, high slip tendency and dilation tendency values for this fault segment indicate a locus of increased tectonic activity. For both sub-model areas the analyses of the slip tendency and dilation tendency cannot provide constraints on the seismicity of individual fault surfaces. It is however interesting to note that the few seismically active faults in the central and northern URG are characterised by significantly lower slip tendency and dilation tendency values than other faults that show evidence of large displacements over geological time spans.

SAMENVATTING (SUMMARY IN DUTCH)

De Boven Rijndal Slenk (BRS) is het centrale segment van een Cenozoiisch slenksysteem in het voorland van de Alpen. De BRS is ontstaan tijdens de vorming van de Alpen. Het ontstaan van de BRS wordt tevens door een signifikante opwelling van de mantel vergezeld. In het huidige spanningsveld wordt de BRS gereactiveerd in de vorm van een sinistraale *shearzone*. De reaktivatie wordt gekenmerkt door relatief langzame oppervlakte deformatie, een regionaal verhoogde kans op seismische aktiviteit en een signifikant merkbare oppervlakte daling (0.1 tot 0.2 mm/jaar geologisch, 1 mm/jaar geodetisch berekende ratios). De factoren die tot de huidige tektonische aktiviteit bijgedragen hebben zijn tot nog toe niet volledig begrepen. Informatie van de spanningstoestand in de aardkorst zijn slechts in weinige gebieden voorhanden. De seimische aktiviteit in de regio is niet beperkt tot de slenkrandbreuken, noch zijn er weinig seismisch aktieve breuken bekend. Voor het grootste deel zijn breuken in de BRS niet bestudeerd en is weinig bekend met betrekking tot hun Kwartaire aktiviteit of de mogelijkheid tot reaktivatie in het huidige spanningsveld. Het doel van dit proefschrift is het onderzoeken en nabootsen van breuken in het huidige spanningsveld met behulp van de eindige elementen methode.

De eindige elementen methode wordt gebruikt voor het inschatten van de kinematische verhoudingen en de in-situ spanningstoestand in een geologisch systeem. De methode wordt op verschillende schalen gebruikt, zowel grootschalig (bv. de aardkorst) als kleinschalig (bv. tunnelboringen). In dit proefschrift worden verschillende schalen gebruikt om het huidige spanningsveld in een regionaal model, dat de gehele BRS omvat, na te bootsen. De in dit regionale model geobserveerde vervormingen en spanningen, worden vervolgens gebruikt als uitgangspunt die twee kleinere en gedetailleerdere modellen van het centrale deel en noorden van de BRS aandrijven. De kleinere en meer gedetailleerde modellen bevatten een groot aantal breuken. Voor het onderzoek van het in dit proefschrift gepresenteerde materiaal zijn verschillende commerciële software programmas, zoals gOcad®, HyperMesh® en ABAQUS™ gebruikt. Er is een methode ontwikkeld welke geometrisch complexe modellen kan genereren en numeriek oplossen.

Verschillende gegevens zoals bv. de geometrische struktuur, materiaal eigenschappen en randvoorwaarden zijn geïntegreerd in dit proefschrift om een model te bouwen en numeriek op te lossen. Het regionale model bestaat uit drie lagen: de bovenkorst, de onderkorst en de lithosferische mantel. De varierende dikten van deze gesteentelagen resulteert in laterale dichtheidsvariaties, welke leiden tot extra spanningen in het regionale model. Daarnaast is de complexe Varistische structuur van de aardkorst nagebootst door middel van dichtheids variaties. De onderkorst en lithosferische mantel zijn gebruikt als extra randvoorwaarden om de bovenkorst aan te drijven. De slenkrandbreuken zijn in het model opgenomen als zogenaamde 'kontaktoppervlakken' die de natuurlijke condities van een breuk zo goed mogelijk nabootsen.

Om het regionale model numeriek op te lossen zijn geschikte randvoorwaarden noodzakelijk. Dit proefschrift richt zich op het vinden van de best passende randvoorwaarden door middel van het testen van verschillende bewegingscondities aan de modelranden. De zwaartekracht werkt op elk element in de richting van het middelpunt van de aarde. Om de spanningstoestand en de kinematische bewegingen in het onderzochte tijdsinterval te simuleren, zijn passende randvoorwaarden alleen niet toereikend. Een correcte initiële spanningstoestand welke de geometrische complexiteit en de geologische geschiedenis weerspiegeld is essentieel. In dit proefschrift zijn daartoe modellen gebruikt welke de ronding van de aarde honoreren om het effect van elastische samendrukking van gesteenten correct na te bootsen. Naast deze modelaannames is in het gehele model van een hydrostatische poriëndruk uitgegaan en van de uitgangs aanwezige tektonische spanningen. De best passende initiële spanningstoestand is na meerdere modelfasen van o.a. zwaartekracht en verschillende grenscondities bereikt. In de laatste modelfase is de nagebootste spanning van het regionale model door middel van in-situ gegevens gekalibreerd. De gemodelleerde bewegingen en vervormingen van het gekalibreerde regionale model zijn vervolgens als grenscondities toegepast op de kleinere en meer gedetailleerde modellen. De resultaten van de kleinere modellen worden vergeleken en gekalibreerd met relatieve bewegingen geobserveerd aan het aardoppervlak en gemeten referentiegegevens.

De resultaten van het regionale model, met lateraal variërende aardkorstdikte en geometrische structuren, hebben een duidelijk effect op het huidige spanningsveld en de recente bewegingen. De resultaten van het regionale model zijn sterk afhankelijkheid van de aangenomen opwelling van de top van de onderkorst en litosferische mantel. De modelresultaten geven duidelijk aan dat de huidige BRS zich in een transtensioneel spanningsveld bevindt en daaraan uitdrukking geeft in de vorm van een sinistraal horizontaal verschuiving-systeem. Het centrale segment uit zich als een *restraining bend*, waarbij de breuk en seismische aktiviteit wisselt tussen de oostenlijke randbreuk en de westelijke randbreuk.

De resultaten van het eerste kleinschalige model duiden erop dat dit segment in het huidige spanningsveld meer aan compressie onderhevig is dan het noordelijke en zuidelijke segment. Het model voorspelt verschillende deformatie stijlen: zowel extensie en compressie breuken. Wanneer seismisch aktieve breuken vergeleken worden met de resultaten van de gemodelleerde breukreactivatie is de conclusie dat destructieve aardbevingen in deze regio voornamelijk langs breuken met een lage reactivatietendens kunnen voorkomen.

De resultaten van het tweede kleinschalige model van het noordelijke deel van de BRS laten zien dat dit segment zich kenmerkt als een releasing bend. Ten opzichte van het huidige spanningsveld wordt de laterale compressie relatief verminderd door een verandering in orientatie van de grensrandbreuk. De resultaten van de breuktendensanalyse en de dilatatieanalyse voor dit breuksegment duiden op een verhoogde tectonische activiteit. Voor beide kleinschalige modellen bieden de analyses geen uitkomst met betrekking tot de seismische activiteit van individuele breuken. Het is echter interessant te vermelden dat de paar seismisch actieve breuken in het centrale en noordelijke deel van de BRS worden gekenmerkt door significant lagere slip tendens en dilatatie tendens waarden dan andere breuken die bewijs tonen voor groot verzet door de geologische tijd heen.

ZUSAMMENFASSUNG (SUMMARY IN GERMAN)

Der Oberrheingraben (ORG) ist das zentrale Segment eines känozoischen Grabensystems im Vorland der Alpen. Der ORG entwickelte sich im Rahmen der Alpenentstehung. Die Enstehung des ORG war von einer signifikanten Hebung der Moho begleitet. Zurzeit wird der ORG als sinistrale Scherzone im regionalen Spannungsfeld reaktiviert. Diese Reaktivierung führt zu einer relativ langsamen tektonischen Deformation, begleitet von regional erhöhter Seismizität und signifikanter lokaler Subsidenz (0.1 bis 0.2 mm/a geologisch, 1 mm/a geodätisch ermittelte Rate). Die Faktoren, die zur heutigen tektonischen Aktivität beitragen, sind bisher noch nicht vollständig verstanden. Informationen zum krustalen Spannungszustand sind nur für wenige Orte erhältlich. Die Seismizität in der Region ist nicht auf die Grabenrandstörungen begrenzt und nur wenige seismisch aktive Störungen sind bekannt. Für die Mehrheit der Störungen im ORG gibt es bislang keine Untersuchungen zu deren quartärer Aktivität oder deren Reaktivierungspotenzial im rezenten Spannungsfeld. Ziel dieser Arbeit ist es daher, das Verhalten der Störungen im rezenten Spannungsfeld mithilfe der Finiten Element Methode zu simulieren.

Die Finite Element Methode kann zur Abschätzung des kinematischen Verhaltens und des in-situ Spannungszustandes in einem geologischen System genutzt werden. Die Methode ist auf verschiedenen Skalen anwendbar und mit ihr lassen sich sowohl großräumige (z.B. krustale Platten) als auch lokale (z.B. Tunnelbau) Fragestellungen beantworten. In dieser Arbeit wurden verschiedenskalige Modelle benutzt. Hierzu wurde das regionale Spannungsfeld mithilfe eines regionalen Modells simuliert, das den gesamten ORG beinhaltet. Versatzbeträge, die mithilfe dieses Modells ermittelt wurden, wurden in einem zweiten Schritt als Randbedingungen für höher aufgelöste Modelle des mittleren und nördlichen Grabensegmentes verwendet. Diese Detailmodelle beinhalten eine grosse Anzahl an Störungen. Im Rahmen dieser Arbeit wurde unter Verwendung kommerzieller Software (gOcad®, HyperMesh®, ABAQUS™) eine Methode entwickelt, mit deren Hilfe geometrisch komplexe Modelle generiert und gelöst werden können.

Verschiedene Datensätze wie z.B. die geometrische Struktur, Materialeigenschaften und Randbedingungen müssen integriert werden, um ein solches Modell zu konstruieren und zu lösen. Für das regionale Modell musste eine Geometrie definiert werden, die es erlaubt das regionale Spannungsfeld und das kinematische Verhalten des ORG zu simulieren. Das regionale Model beinhaltet drei Lithosphärenschichten (Oberkruste, Unterkruste und lithosphärischer Mantel). Die varierenden Mächtigkeiten dieser Schichten führen zu lateralen Dichteunterschieden, die zusätzliche Spannungen im Modellraum generieren. Darüber hinaus wurde die komplexe variskische Struktur der Kruste mithilfe von lateralen Dichteunterschieden simuliert. Die Unterkruste und der obere Mantel wurden zusätzlich dazu benutzt, eine weitere Randbedingung an der Basis der Oberkruste aufzugeben. Das Modell beinhaltet die Grabenrandstörungen als Kontaktflächen.

Um das regionale Modell lösen zu können, müssen geeignete Randbedingungen gewählt werden. Im Rahmen dieser Arbeit wurden Verschiebungsrandbedingungen an den Seiten und der Basis der Modelle angewandt. Darüberhinaus wirkt die Gravitation auf jedes Element in Richtung des Erdmittelpunktes. Um den Spannungszustand und das kinematische Verhalten eines tektonischen Systems zu simulieren, genügt es nicht, geeignete Randbedingungen für den untersuchten Zeitraum zu definieren. Vielmehr muss ein geeigneter Ausgangsspannungszustand für das Modell definiert werden, der der geometrischen Komplexität und der geologischen Geschichte entspricht. Im Rahmen dieser Arbeit wurden sphärische Modellgeometrien verwendet, um den Effekt der elastischen Kompaktion im Modellraum zu simulieren. Darüberhinaus wurde angenommen, dass im gesamten Modellraum hydrostatischer Porendruck herrscht und tektonische Spannungen wirken. Der Ausgangs-spannungszustand wurde in mehreren Schritten erreicht, die unter anderem die Anwendung der Gravitation und Verschiebungsrandbedingungen beinhalten. Im letzten Schritt wurde der simulierte Spannungszustand des regionalen Modells mithilfe von in-situ Spannungsdaten kalibriert. Die ermittelten Verschiebungsbeträge aus dem regionalen Modell wurden nach erfolgter Kalibrierung als Verschiebungs-randbedingungen auf die Modellränder der lokal skaligen Modelle aufgegeben. Um die Modellergebnisse dieser Modelle evaluieren zu können, wurden die relativen Hebungsbeträge und Störungsbewegungen mit Referenzdaten verglichen.

Die Ergebnisse des regionalen Modells zeigen einen grossen Einfluss der unterschiedlichen Krustendicke und der geometrischen Struktur im Modell auf den in-situ Spannungszustand und das rezente kinematische Verhalten. Desweiteren zeigt das Modell eine starke Abhängigkeit von der angenommenen Hebung der Moho an. Die Modellergebnisse legen nahe, dass sich der ORG zurzeit in einem transtensionalen Spannungsfeld als sinistrales Horizontalverschiebungssystem verhält. Das zentrale Grabensegment bildet einen *„restraining bend"*. Hierbei wechselt die

Aktivität mehrfach von der östlichen auf die westliche Grabenrandstörung.

Die Ergebnisse des ersten lokal skaligen Modells vom mittleren ORG-Segment zeigen, dass dieses Segment im heutigen Spannungsfeld kompressiver ist als das nördliche und südliche Segment. Das Model prognostiziert unterschiedliche Deformationsarten, sowohl Extension als auch Kompression, für die krustalen Störungen. Ein Vergleich der seismogenen Störungen mit den Ergebnissen der Gleittendenzanalyse zeigt, dass Schadenbeben in dieser Region möglicherweise an Störungen gebunden sind, die eine geringe Gleittendenz aufweisen.

Die Ergebnisse des zweiten lokal skaligen Modells vom nördlichen Segment des ORG zeigen, dass dieses Segment im heutigen Spannungsfeld einen „releasing bend" bildet. Die laterale Kompression in diesem Grabenabschnitt ist durch einen Wechsel in der Orientierung des ORG relativ zum Spannungsfeld erniedrigt. Das Segment der östlichen Grabenrandstörung, das den größten Vertikalversatz im Modell zeigt, fällt mit der Lokation des ausgeprägtesten quartären Depozenters zusammen. Die Ergebnisse der Gleittendenz- und Dilatationstendenzanalyse für dieses Störungssegment zeigen außerdem erhöhte tektonische Aktivität an. Für beide lokal skalige Modelle lassen die Ergebnisse der Gleittendenz- und Dilatationstendenzanalyse keine Rückschlüsse über die potenzielle Seismizität einzelner Störungen zu. Eine interessante Beobachtung ist jedoch, dass die wenigen seismisch aktiven Störungen im zentralen und nördlichen ORG durch signifikant niedrigere Gleit- und Dilatationstendenz charakterisiert sind als Störungen, die große Versatzbeträge über geologische Zeiträume aufweisen.

INTRODUCTION AND THESIS OUTLINE

1.1 Problem statement

The knowledge of the three dimensional (3D) in-situ state of stress of a geological system is essential both for many geomechanical problems and for the understanding of geodynamic processes. For example, knowledge of the in-situ stress state can be applied to assess the stability of boreholes during drilling and for the optimisation of the production process in hydrocarbon industry. Furthermore, the knowledge of the in-situ state of stress can be used to investigate fault reactivation and fracture generation. Within the Earth's crust the 3D state of stress defines the tectonic regimes (see section 2.1.2), under which the fracturing and faulting of rocks initiates and determines the most seismically active and hazardous regions.

The in-situ state of stress can be derived by several methods (e.g. overcoring, hydraulic fracturing, see section 2.1). These observations however only provide information on the state of stress in the vicinity of the measurement locations and are may be related to local crustal heterogeneities. In order to obtain a volumetric approximation of the 3D state of stress within a geological volume the finite element method (FEM) is a useful tool. By using an approximation of the systems geometry and the spatial distribution of material properties, the 3D state of stress can be extrapolated from locally derived in-situ stress observations. Therefore, using the approach of the FEM to simulate the state of stress within a geological volume, it is essential to calibrate the model with in-situ stress data.

In this study, the region of the Upper Rhine Graben (URG) is used to simulate the contemporary state of stress using the FEM. The URG is amongst the most extensively studied intra-continental graben systems. Therefore, an extensive set of calibration and benchmarking data exists, obtained by several Earth Science disciplines (e.g. structural geology, sedimentology, seismology and geomorphology; ILLIES & FUCHS, 1974; BEHRMANN et al., 2005; CLOETINGH & CORNU, 2005; CLOETINGH et al., 2007). The diversity of these different types of reference data provides constraints for the tectonic behaviour on a range of geological time-scales. In addition, in the URG area reference data from both industrial exploration (hydrocarbon and geothermal) and geoscientific research are available. For this reason, the URG represents an excellent test area for the modelling approach presented in this study. The aim of this study is:

- To define a modelling strategy, that can provide a reliable and repeatable approximation of the 3D state of stress within the crust.

- To develop a series of crustal-scale finite element models that approximate the 3D state of stress within the URG. These models are applied to scales of the entire URG and its northern and central segments.

- To investigate the contemporary kinematics of the URG based on modelling results such as predicted displacements along implemented faults and predicted surface motions.

- To evaluate the reactivation potential of pre-existing faults associated with the graben structure under the approximated 3D state of stress. Furthermore, based on the evaluation of the fault reactivation potential, a possible contribution of 3 dimensional mechanical earth modelling (MEM) to earthquake hazard assessment is investigated.

In the following sections, an overview of the geological evolution, the crustal structure, the observed fault kinematics and the present-day stress field of the modelling area is given. The last part of this chapter comprises a brief thesis outline including a summary of each thesis chapter.

1.2   Geological setting of the Upper Rhine Graben area

The Upper Rhine Graben (URG) forms the central part of the European Cenozoic Rift System (ECRIS; ZIEGLER, 1992, 1994; PRODEHL et al., 1995), which is a system of rift related sedimentary basins connecting the western Mediterranean with the North Sea over a distance of about 1000 km (Figure 1.1). Development of the ECRIS is linked to collisional deformation in the Alpine foreland accompanied by mantle uplift (ILLIES, 1974B; MONNINGER, 1985; ZIEGLER, 1990, 1994). The sedimentary record (as revealed from boreholes and seismic studies) shows that rifting processes within the ECRIS began in the Middle Eocene (e.g. WITTMANN, 1955; SITTLER & SONNE, 1971; ILLIES, 1977; SISSINGH, 1998; LUTZ & CLEINTUAR, 1999; HINSKEN et al., 2007).

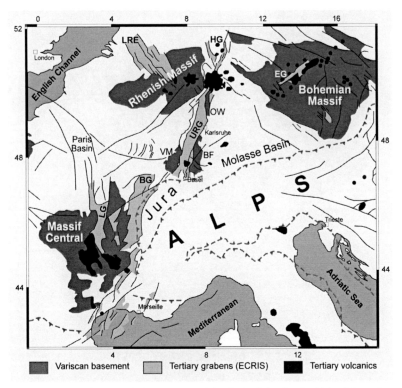

Figure 1.1: Map of the ECRIS in the European Alpine foreland showing the Alpine deformation front (dashed light grey line), Cenozoic rift related sedimentary basins and volcanic centres, Cenozoic fault systems (black lines) and Variscan Massifs. LRE, Lower Rhine Embayment; HG, Hessian grabens; EG, Eger Graben; OW, Odenwald; VM, Vosges Mountains; BF, Black Forest; BG, Bresse Graben; LG, Limagne Graben (after DÈZES et al., 2004).

The URG is a NNE-SSW trending graben approximately 300 km long and 35 to 40 km wide. It is bounded in the north by the Hunsrück Taunus Boundary Fault (HTBF) and the Miocene Vogelsberg volcano and in the south by the Rhine Saône Transfer Zone (RSTZ) leading towards the Bresse Graben (BG) and the frontal thrusts of the Jura Mountains (Figure 1.2). The graben has undergone a complex evolution involving reactivation of inherited structures and multiple changes in regional kinematics (e.g. SCHUMACHER, 2002; DÈZES et al., 2004). The crustal extension, perpendicular to the rift axis that occurred during the graben evolution is estimated to be in the order of 5 to 7 km (ILLIES, 1965; ZIEGLER, 1994; HINSKEN et al., 2007).

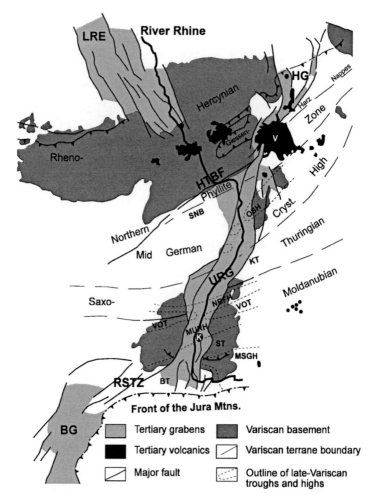

Figure 1.2: Geological overview of the URG area and adjacent tectonic units. The crustal structure of the URG area includes Tertiary grabens and volcanic complexes, Variscan and Late Variscan structural domains (BOIGK & SCHÖNEICH, 1970; FRANKE, 1989; ZIEGLER et al., 2004). LRE, Lower Rhine Embayment; HG, Hessian grabens; BG, Bresse Graben; V, Vogelsberg volcano; K, Kaiserstuhl volcano; HTBF, Hunsrück Taunus Border Fault; RSTZ, Rhine Saône Transfer Zone; SNB, Saar Nahe Basin; OSH, Odenwald Spessart High; KT, Kraichgau Trough; NBFH, Northern Black Forest High; VOT, Villé Offenburg Trough; MURH, Main Upper Rhine High; ST, Schramberg Trough; MSGH, Main Southern German High; BT, Burgundy Trough.

Within the URG a number of sub-basins exist, that reflect the migration of the main depocentres (SCHUMACHER, 2002) in response to changes in the regional and local tectonic and magmatic conditions. At present, the URG shows low to moderate seismicity (e.g. AHORNER, 1983; BONJER, 1997) but a significant amount of subsidence (0.1 to 0.2 mm/a geological and 1 mm/a geodetical rate) in the recent depocentres (e.g. PFLUG, 1982; MONNINGER, 1985).

1.3 Tectonic evolution of the Upper Rhine Graben

1.3.1 Pre-rift setting of the URG area

During the Cenozoic rifting process, reactivation of Variscan and Late Variscan basement structures had a strong control on the developing geometry of the URG an (e.g. SCHUMACHER, 2002). The URG developed in an area characterised by a complex pre-Tertiary crustal architecture (Figure 1.2). The URG cuts across Several Variscan terranes and terrane boundaries (ZIEGLER et al., 2004), and the following Variscan trends are observed in the URG fault pattern: (1) the NNE-SSW orientation of the URG, the so-called Rhenish trend, (2) the NE-SW to ENE-WSW oriented Erzgebirgian trend and (3) the NW-SE oriented Hercynian trend. In the URG area, the NE-SW to ENE-WSW striking Variscan terranes (e.g. Saxothuringian domain) were overprinted by Late Variscan wrench tectonics, during which Variscan fault systems were reactivated to form a system of sub-parallel Permo-Carboniferous troughs and highs (e.g. ZIEGLER, 1990; ZIEGLER et al., 2004 Figure. 1.3). This Late Variscan structural fragmentation of the Central European crust involved the wrech induced transtensional collapse of the Variscan orogen during Permo-Carboniferous times (Stephanien to Autunian; e.g. LAUBSCHER, 1986; ZIEGLER, 1990; EDEL & WEBER, 1995; ZIEGLER et al., 2004; ZIEGLER & DÈZES, 2006).

During late Permian to Jurassic times the URG area was characterised by thermal subsidence and the development of a new system of intra-cratonic basins (ZIEGLER et al., 2004; ZIEGLER & DÈZES 2005). Triassic to Late Jurassic sediments, in general thickening to the north, rest discordantly on the Permo-Carboniferous sediments (PFLUG, 1982). At present, Triassic Buntsandstein is exposed on the graben shoulders in the Pfälzer Wald, the northern Vosges Mountains and the northern Black Forest. The Kraichgau, north of the Black Forest, is dominated by the exposure of limestones of the Muschelkalk. Outcrops of Jurassic sediments are only found on the eastern shoulder in the Kraichgau and on the western shoulder in the Zabern Trough. Furthermore, Jurassic sediments are exposed within the graben in a small zone east of the Vosges Mountains, and in a zone between Freiburg and Basel.

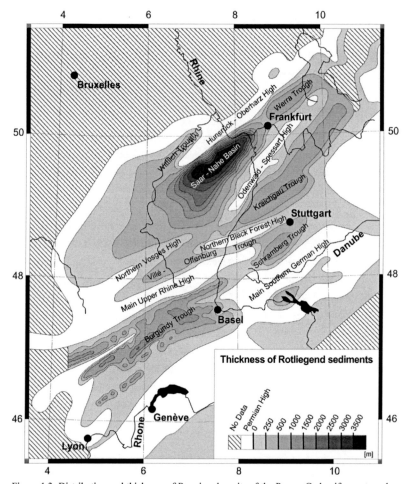

Figure 1.3: Distribution and thickness of Permian deposits of the Permo-Carboniferous troughs in the URG area (after BOIGK & SCHÖNEICH, 1970). The distribution and thickness variation indicate the complex crustal structure after the Variscan orogeny.

A major sedimentary hiatus occurred in Cretaceous and Palaeocene times, often attributed to the Late Jurassic and Cretaceous uplift of the so-called Rhenish Shield, giving rise to uncertainty as to whether Cretaceous sediments were deposited in the URG region or not (CLOOS, 1939; ILLIES, 1975; ZIEGLER, 1990; WALTER, 1995). As a result of the missing Cretaceous deposits, the Eocene and younger synrift sediments within the URG discordantly rest on Mesozoic sediments throughout the graben. In the northern URG, the Mesozoic sediments comprise Triassic units while in the southern URG they also comprise Jurassic sediments. During Cretaceous and Palaeocene times collision-related compressional deformation affected the relatively weak Variscan crust of Central Europe in response of the closing of the Penninic Ocean that had separated the European Plate from the Adriatic Indentor (DÈZES et al., 2004). The resulting intraplate compression caused large-scale inversion of basins and lithospheric folding in the foreland of the evolving Alps (e.g. ZIEGLER 1992; ZIEGLER et al., 1998; MICHON et al., 2003). At this time, an initial phase of pre-rift volcanism affected the Bohemian Massif, the Rhenish Massif and the southern part of the Vosges-Black Forest Massif. The beginning of rift related magmatism (primitive alkali basaltic rocks) in the URG area was dated by KELLER et al. (2002) to be of Palaeocene age.

1.3.2 Synrift evolution of the URG

The onset of rifting in the URG area is dated by sediments of Lutetian age (middle Eocene; e.g. ILLIES, 1977; SISSINGH, 1998; LUTZ & CLEINTUAR, 1999; SCHUMACHER, 2002; HINSKEN et al., 2007). During the formation of the URG, several reactivation phases have affected the Variscan and Late Variscan structures (e.g. SCHUMACHER, 2002), in particular the HTBF at the northern end of the graben (Figure 1.2).

The total thickness of Tertiary to Quaternary sediments within the URG exceeds 3200 m in an elongated depocentre situated along the eastern graben border in the northern URG at Mannheim (DOEBL & OLBRECHT, 1974). In the southern URG, a smaller depocentre south of the Kaiserstuhl exists, filled by a sequence of 2500 m thick sediments (Figure 1.4).

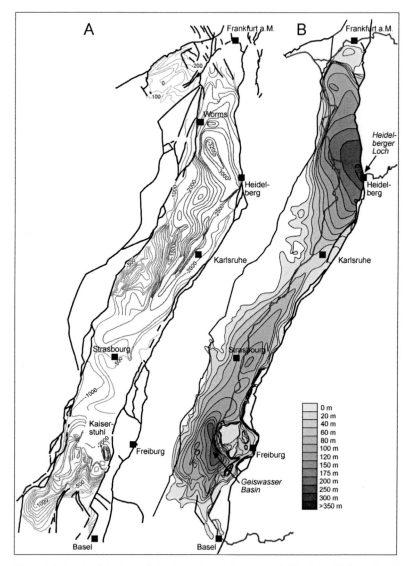

Figure 1.4: Depth of the base of the Cenozoic and Quaternary infill of the URG. A) Depth contours of the base of the Tertiary deposits (relative to the surface) after DOEBL & OLBRECHT (1974). B) Depth contours of the base of the Quaternary deposits, after BARTZ (1974) and HAIMBERGER et al. (2005). Note the two distinct areas of increased subsidence and deposition in the north (i.e. Heidelberger Loch) and the south (i.e. Geiswasser Basin).

The graben fill consists of a sequence of lacustrine and fluvial sediments occasionally interrupted by sediments relating to marine transgressions (see compilations in PFLUG, 1982; SCHUMACHER, 2002; DERER et al., 2003; BERGER et al., 2005). The Mainz Basin, located at the north-western end of the URG, corresponds to a shallow sub-basin that formed as part of the URG during the early stages of rifting and evolved independently by Late Oligocene times. It should be noted that the stratigraphy is the same for both the URG and the Mainz Basin until the Early Miocene (Burdigalian Upper Hydrobia Beds; BERGER et al., 2005). Subsidence and sedimentation started in the Lutetian in local basins throughout the region (e.g. ILLIES, 1977; SISSINGH, 1998; LUTZ & CLEINTUAR, 1999; SCHUMACHER, 2002). The oldest deposits preserved in the URG are siderolitic formations and residual clays, documenting lateritic and karstic weathering of Mesozoic platform carbonates, locally associated with lacustrine shales, limestones and lignites (SCHMIDT-KITTLER, 1987; SCHUMACHER, 2002).

By Priabonian times (Late Eocene), depositional conditions re-established with areas in the southern and central URG, the Mulhouse and Strasbourg (Zorn) sub-basins continued to subside (SCHUMACHER, 2002). In these sub-basins a terrestrial, mainly lacustrine, series was deposited. Coarse conglomerates occurring along the basin margins indicate increased erosion due to uplift of the rift flanks. A temporary marine connection between the Mulhouse Basin and the Molasse Basin is possible (SITTLER, 1965; SISSINGH, 1998). Priabonian sediments rest unconformable on subcropping Mesozoic and Early to Middle Eocene strata (SISSINGH, 1998; SCHUMACHER, 2002). More than 800 m of Priabonian sediments were deposited in the two sub-basins (PFLUG, 1982), which were superimposed on Late Variscan troughs that were probably being reactivated (SCHUMACHER, 2002).

Deposition continued in the southern and central URG during the Early Oligocene (Rupelian) and continuously propagated northwards. Lower Rupelian sediments were deposited as far north as in the Hessian Strait in the area of Kassel (SCHUMACHER, 2002). Depositional conditions remained dominantly lacustrine to terrestrial (SISSINGH, 1998) with large volumes of conglomerates deposited at the margins of the rift basin. This is interpreted as indicating an increased relief gradient (DURINGER, 1988). A maximum of ca. 800 m Early Rupelian Pechelbronn Beds (middle part of the URG) and Upper Salt Formation (southern part of the URG) was deposited in the basin. Eventually, during the Upper Rupelian, the URG acted as a marine channel connecting the Lower Marine Molasse Basin in the south to the North Sea Basin in the north (GRIMM, 1994; BERGER, 1996; REICHENBACHER, 2000). Uniform deposition of up to 400 m of

Upper Rupelian sediments occurred throughout the URG during this period.

The marine connection of the URG with the North Sea Basin was interrupted in the Late Oligocene (Chattian; ANDERSON, 1961; GRAMANN, 1966; KADOLSKY, 1988) and the URG was again dominated by a fluvio-lacustrine brackish to freshwater environment (SCHUMACHER, 2002). By this time, the main depocentre within the URG at this time was located in the central part of the graben (Strasbourg sub-basin). More than 400 m of the Bunte Niederrödener Schichten formation were deposited at this sub-basin, whereas the northern and southern part of the graben was temporarily affected by uplift and erosion. Only 20 m of the Bunte Niederrödener Schichten formation was deposited in the Mainz Basin (ROTHAUSEN & SONNE, 1984). In general, deformation in the middle and northern part of the graben was more complex than in the south as documented by synsedimentary faulting activity, whereas subsidence was more homogeneously distributed during the middle Oligocene (PFLUG, 1982).

The Oligocene - Miocene boundary is marked by a major change in rift kinematics (e.g. SCHUMACHER, 2002). The depocentre within the URG shifted northward whilst the southern and central parts of the URG experienced uplift (e.g. SCHUMACHER, 2002; Figure 1.5). Increased sediment influx from the Vosges Mountains and the Black Forest indicate relatively rapid uplift of the graben shoulders in the south of the URG (LAUBSCHER, 1992; KÄLIN, 1997) which was accompanied by activity of the Kaiserstuhl volcano (7 – 18 Ma ago; e.g. DÈZES et al., 2005). This rapid uplift of the southern and central URG was probably induced by lithospheric folding as a result of Alpine collisional forces. Contemporaneously, progressive thermal doming of the Rhenish Massif occurred during Middle to Late Miocene times (FUCHS, et al. 1983) accompanied mayor volcanic activity of the Vogelsberg (18 – 16 Ma; BOGAARD & WÖRNER, 2003) in the north of the URG. The uplift of the Rhenish Massif also involved the Mainz Basin, placing this region in a marginal location to the URG. Since Early Miocene times the northern URG and the Mainz Basin are thought to be isolated from the surrounding marine basins (GEBHARD, 2003).

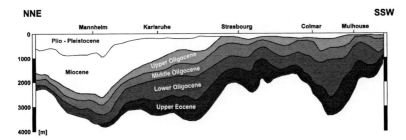

Figure 1.5: Schematic along strike profile of the sediment infill of the URG (modified after PFLUG, 1982). Note the transfer of the locus of greatest subsidence to the northern URG at the Oligocene – Miocene boundary.

The maximum accumulation of Miocene and Pliocene fluvial sediments occurred in the area of the Heidelberger Loch in the northern URG (1500 m of Miocene, 760 m of Pliocene sediments; PFLUG, 1982). The Heidelberger Loch remained the dominant depocentre throughout the Quaternary (> 380 m of Quaternary alluvial deposits; BARTZ, 1974). In addition, a second Quaternary depocentre, the Geiswasser Becken, evolved in the southern URG, southwest of the Freiburg Embayment (e.g. SCHUMACHER, 2002, Figure 1.4 B). During early Quaternary the Mulhouse Swell and the Sundgau area, both located at the southern end of the graben, were uplifted (SCHUMACHER, 2002) and deformation of the Sundgau Gravels occurred (THÉOBALD, 1934). The river Rhine is now draining through the entire URG to the north (VILLINGER, 1998).

Based on the sedimentary record and sub-basin geometries SCHUMACHER (2002) compiled a time scale for the development of the URG rift system (Figure 1.6). This compilation includes proposed fault kinematics under different stress conditions during multiple tectonic phases of the URG development. The number of tectonic phases is still a matter of debate (see discussion in SCHWARZ, 2006). However, it is commonly agreed that the rifting in middle Eocene times initiated under an E-W extensional stress field. During the Oligocene, pre-existing weakness zones were reactivated in an overall extensional to transtensional stress field with E-W to NW-SE oriented extension. As a result of this stress field the entire graben subsided as documented by sedimentation (Figure 1.5). The stress field reoriented at the beginning of the Miocene and changed to sinistral transtension, persisting until the present-day. During this last phase of NW-SE oriented compression and NE-SW oriented extension, the URG system was reactivated by sinistral shearing (ILLIES, 1975, SCHUMACHER, 2002, MICHON et al., 2003, DÈZES et al., 2004).

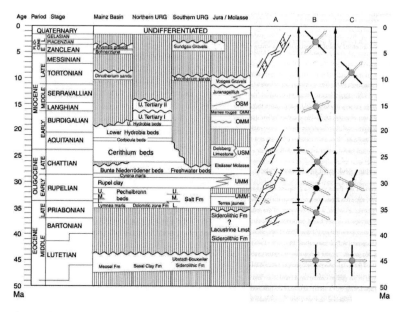

Figure 1.6: Stratigraphic table for Cenozoic sediments in the URG and adjacent tectonic units combined with kinematic models (column A) for different phases of the URG evolution (SCHUMACHER, 2002 and references therein). Additionally, palaeostress orientations after VILLEMIN & BERGERAT (1987; column B) and MICHON et al. (2003; column C) are shown (LOPES CARDOZO, 2004; black, maximum stress component; shaded, intermediate stress component; white minimum stress component).

Contemporary geological processes within the URG area are still influenced by Variscan to Late Variscan and Alpine structures. This becomes evident from the present-day topography of the crust / mantle boundary (Moho). In the northwest of the URG the Moho shows a distinct topography associated with Late Variscan highs and troughs. A distinct high of the Moho (Vosges Black Forest Dome; VBFD), centred in the southern URG, is associated with the locus of Miocene to Pliocene uplift of the Vosges Black Forest Massif (Figure 1.7).

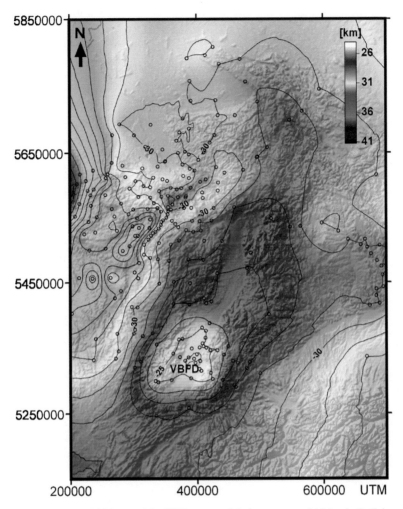

Figure 1.7: Crustal thickness of the URG area overlain by contours of Moho depth (1 km spacing). The topography of the Moho was obtained from seismic reflection and refraction data (after BARTH, 2002). Data points are indicated by dots. VBFD = Vosges Black Forest Dome.

In addition to the Moho topography, both the Variscan structural domains and their Late Variscan break up are visible in the present-day surface geology and topography of the graben shoulders (Figure 1.1 & Figure 1.2). Exhumed Variscan crust (Hunsrück, Taunus, Odenwald, Vosges Mountains and Black Forest) coincide with the highest present-day topography whilst Cenozoic troughs and topographic low areas coincide with Late Variscan troughs (e.g. Kraichgau Trough).

1.3.3 Crustal structure of the Upper Rhine Graben

The deep crustal structure of the URG area was investigated using two seismic lines shot across the URG in 1988 by German and French Deep Reflection Seismic Consortiae (ECORS and DEKORP; Brun et al., 1991; Wenzel et al., 1991), which resolve the structure of the entire crust and parts of the upper mantle. The two seismic lines cross the URG in the northern segment (DEKORP9N; Figure 1.8 a) and the southern segment (DEKORP9S; Fig. 1.8 b) reveal an upper crustal thickness of approximately 16 to 18 km and a total crustal thickness of 24 to 27 km. The dome shaped structure in the southern segment of the URG resolved by the southern profile is referred herein as the Vosges Black Forest Dome (VBFD; alternatively termed as the Vosges-Black Forest Arch, e.g. Dézes et al., 2004). The VBFD represents the thinnest continental crust north of the Alps (Dézes & Ziegler, 2002; Dézes et al., 2004; Figure 1.7). The two deep seismic profiles reveal the traces of the geometry of the graben bounding faults relatively well for the shallower parts of the upper crust, whereas the deeper geometry is less constrained (Fig. 1.8). It is assumed that the graben bounding faults extend to the upper crust / lower crust boundary (i.e. brittle ductile transition zone; BDTZ). Numerous seismic surveys undertaken for hydrocarbon exploration in the 1960s and 1970s reveal that the upper crust in the URG area is highly faulted. These second order faults extend to a maximum depth of approximately 10 km into the basement underneath the graben (Brun et al., 1991; Wenzel et al., 1991; Figure. 1.8). The depth of the BDTZ underneath the graben is also constrained by the lower limit of brittle deformation inferred from the observation of seismic events (e.g. Leydecker, 2005a, Bonjer et al., 1984; Plenefisch & Bonjer, 1997). According to Brun (1999) it is generally assumed that ductile deformation occurs within the lower crust of the URG area. As an exception, brittle failure is documented to occur in the southern part of the URG almost at the crust-mantle boundary in the Dinkelberg area east of Basel (Bonjer, 1997).

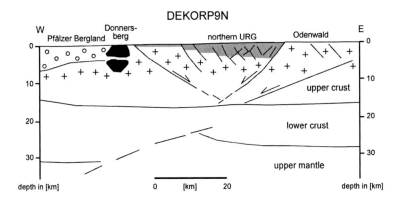

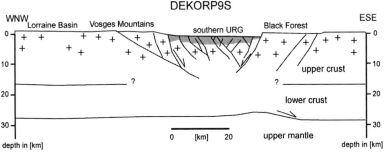

Figure 1.8: Crustal structure of the URG area interpreted from the DEKORP9 seismic reflection profiles (after WENZEL et al., 1991 and BRUN et al., 1991). The existence of low angle dipping shear zones in both profiles (a & b) as well as the lower crustal shear zone in DEKORP9N profile (a) are a matter of ongoing debate (ZEIS et al., 1990; HENK, 1993; SCHWARZ & HENK, 2005). The Cenozoic graben infill is indicated in grey, in the northern profile Permo-Carboniferous volcanic rocks (rhyolites) are indicated in black and Permo-Carboniferous clastic sediments of the Saar-Nahe-Basin are indicated as circles.

1.3.4 Fault kinematics within Upper Rhine Graben area

Discrete fault structures are a prominent feature in the sedimentary infill as well as in shoulder regions of the URG. In the shoulder regions, the fault systems were primarily identified by surficial geological mapping (e.g. ILLIES, 1974A; TIETZE et al., 1979; STAPF, 1988). In contrast, the identification of the intra-graben fault system was primarily conducted on the base of reflection seismic data acquired during the exploration for Oligocene and Miocene hydrocarbon reservoirs (at approximately 1 – 2 km depth). Both methods provide only minor constraints for the more recent (i.e. Pliocene to

Quaternary) deformation history of the URG since little emphasis was placed on mapping faults in economically irrelevant shallower formations. In recent times scientific reflection seismic investigations focused on the structural architecture of Pliocene to Quaternary formations provided new insights to the younger deformation history of the URG which documented displacements in the order of tens of meters within Quaternary sediments (HAIMBERGER et al., 2005; WIRSING et al., 2007). The kinematics of the intra-graben faults is interpreted to be predominantly extensional and total vertical displacements of several hundred to thousands of meters are documented for many faults in 2D seismic profiles (e.g. PFLUG, 1982). Since in 2D seismic profiles horizontal displacements are poorly resolved, the horizontal component of fault movements is generally underestimated. In addition, several authors also suggest significant lateral displacement of specific fault structures (BOSUM & ULLRICH, 1970; MEIER & EISBACHER, 1991; ZIEGLER, 1992; GROSHONG, 1996; LAUBSCHER, 2001). Locally, inversion structures are also documented in the sedimentary infill of the graben (e.g. ILLIES, 1974B). A compilation and detailed review of first (i.e. border faults) and second order (i.e intra graben faults and faults of the shoulder region) fault structures within the URG area is given by PETERS (2007; Figure 1.9 I).

Figure 1.9 (next page): Compilation of first and second order fault structures within the URG area after PETERS (2007; based on ANDRES & SCHAD, 1959; STRAUB, 1962; BEHNKE et al., 1967; BREYER & DOHR, 1967; ILLIES, 1967 & 1974A; TIETZE et al., 1979; STAPF, 1988; DERER et al., 2003). The histograms show the frequency distribution of fault strikes. I) Fault strikes displayed for the entire URG system: bold black lines indicate the graben bounding faults, Active fault segments associated with historical earthquakes are marked in grey (after FRACASSI et al., 2005; LOPES CARDOZO et al., 2005). II) Intra-graben faults. III) Faults of the shoulder region.

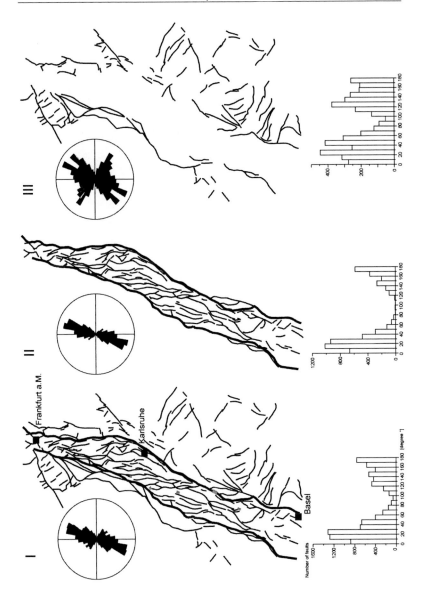

The graben bounding faults and intra-graben faults predominantly strike NNE with subsidiary sets striking N to NW (Fig 1.9 II). The fault pattern within the shoulder region is more complex with main orientations of fault strike of NNE, NE and NW, whereas the maximum at NNE represents graben parallel faults and the NE and NW orientations correspond to Variscan and Late-Variscan fault trends (Fig 1.9 III). As described above, constraints on the recent faulting history of the URG are not available in great detail. ILLIES & GREINER (1978) investigated neotectonic fault movements in the URG area based on 2D seismic sections and concluded that these movements predominantly occur on 170 to 180° striking faults which they interpreted as Riedel shears with left lateral movement. In contrast, interpretations from earthquake fault plane solutions indicate dominantly 150° striking nodal planes for the northern URG and a conjugated set of 20° and 120° striking nodal planes for the southern URG (BONJER et al., 1984). PETERS (2007) identified several fault segments within the northern URG to be active in the Pleistocene using palaeoseismological and geomorphological methods and established faults with documented Pleistocene activity for the URG (Figure 1.10).

Figure 1.10 (next page): Compilation of fault segments within the URG system with documented Pleistocene activity by PETERS (2007). This compilation is based on data by BREYER & DOHR (1967), AHONER & SCHNEIDER (1974), ILLIES (1975), CUSHING et al. (2000), HAIMBERGER et al. (2005) and PETERS (2007). For references of the faults with no age constraint, see figure 1.9. RA=Rastatt.

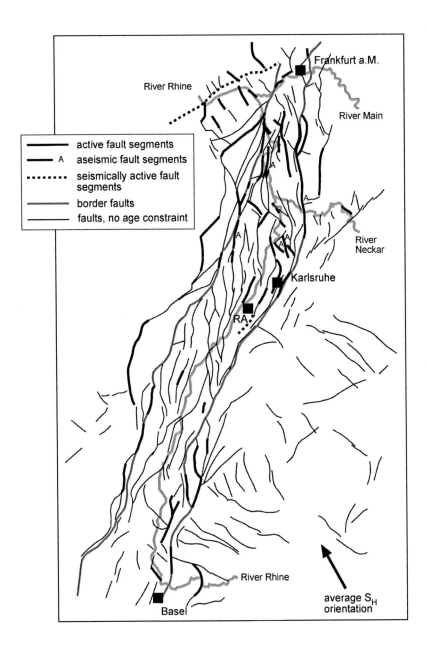

active fault segments

A   aseismic fault segments

seismically active fault segments

border faults

faults, no age constraint

River Rhine

Frankfurt a.M.

River Main

River Neckar

Karlsruhe

RA

River Rhine

Basel

average S_H orientation

1.4 Present-day kinematics, crustal state of stress and seismicity of the URG area

1.4.1 Present-day kinematics

The ongoing African-Eurasian convergence is argued to be the main driving mechanism for the present-day kinematics of the Alps and the northern Alpine foreland (e.g. MÜLLER et al., 1992; PLENEFISCH & BONJER, 1997). Additionally, the kinematics of Central Europe is influenced by a push from the Mid-Atlantic ridge (e.g. GÖLKE & COBLENTZ, 1996). Therefore, the present-day kinematics of the URG is thought to be the consequence of forces associated with both the collision along the southern plate margin and a Mid-Atlantic ridge push (e.g. MÜLLER et al., 1992; GRÜNTHAL & STROHMEYER, 1992; GÖLKE & COBLENTZ, 1996). However, due to the small, relative motions and insufficient accuracy of triangulation data the precise influence of this convergence on the present-day intra-plate kinematics, or even on the kinematics of the Alpine belt, is largely unknown (VIGNY et al., 2002). Several plate motion models predict a NW to NNW (335° to 350°) directed motion of the African plate relative to the European plate in the order of 3 to 8 mm/a in the area of the central Mediterranean (e.g. DeMETS et al., 1990; KREEMER & HOLT, 2001; NOCQUET & CALAIS, 2004). Unfortunately, it remains unclear as to how much of this movement is accommodated by the active mountain belts and how much of it is influencing the present-day kinematics of Central Europe. VIGNY et al. (2002) determined N-S shortening across the Western Alps of less than 1 mm/a, although current GPS measurements of the velocity field still have large uncertainties, particularly in the N-S component (VIGNY et al., 2002). The relative E-W movement within Central Europe is better constrained. NOCQUET & CALAIS (2004) proposed a SW directed movement of the stable part of France (outside the Alps and Jura) relative to Central Europe of less than 0.5 mm/a. This is in agreement with the model of VIGNY et al. (2002), wherein Africa is pushing Western Europe towards the NW, accompanied by a slow dislocation of Western Europe against stable Eastern Europe across the ECRIS and a slow E-W opening of the rift system. Recent measurements by RÓZSA et al. (2005) of permanent and campaign GPS stations located in the URG revealed also horizontal displacements of less than 0.5 – 1.0 mm/a. The short observation time (3 years only) for this study precluded determination of the direction of motion. Further support for these low GPS velocities comes from a modelling study of TESAURO et al. (2005). Using a crustal four-block model of Central Europe, strain rates based on the GPS velocities have been calculated. The results yielded a N to NW directed motion of 0.76 mm/a of the crustal block located to the east and north of the URG and the Lower

Rhine Graben and an E-W motion of 0.51 mm/a of the crustal block located to the west and south of the URG and Lower Rhine Graben respectively.

1.4.2 Present-day stress field

The present-day stress field of the URG area was first investigated in the late 1970s. GREINER & ILLIES (1977) and ILLIES & GREINER (1979) presented in-situ stress determinations, using the so-called door stopper method. These studies indicated a relative homogeneous near surface stress pattern for the URG area and the Molasse Basin (Figure 1.11). In general, ILLIES & GREINER (1979) determined a NW–SE to NNW-SSE orientation of the maximum horizontal stress component (S_H). A disadvantage of the door stopper method is the shallow sampling depth (see chapter 2.1). For the URG area, a maximum sampling depth of 140 m was reached.

In order to achieve a more depth independent model of the URG and adjacent areas, the present-day stress field has been the subject of several seismological studies, using stress tensors inverted from earthquake focal plane solutions (AHORNER et al. 1983; LARROQUE et al., 1987; DELOUIS et al., 1993; PLENEFISCH & BONJER, 1997; HINZEN, 2003). To obtain both the spatial orientation of the two nodal planes, of which one is assumed to be the earthquake rupture plane, and the spatial orientation of the so-called P-, B- and T-axes of an earthquake event, a focal plane solution is calculated. Under the assumption that these three strain axes corresponds to the principal stress axes:

P-axis (pressure)	:	maximum principal stress axis (σ_1)
B-axis (neutral)	:	intermediate principal stress axis (σ_2)
T-axis (tension)	:	minimum principal stress axis (σ_3)

it is possible to infer the intra-plate stress orientations from the inversion of earthquake focal mechanisms (see section 2.1). The quality of the determination of the principal stress axes inferred is strongly dependent on the quality of the earthquake characteristics recorded. For the area of the southern URG, a large set of high quality data is available, since both a dense network of seismometers has been developed and earthquakes occur frequently. Therefore, a precise location and characterisation of the earthquake events in the southern URG is possible. For the northern URG the available data-set is significantly poorer, since earthquakes are less frequent and the monitoring network is less dense. For this reason, seismological studies of the northern

URG also include earthquake data from the more seismically active Rhenish Massif and Lower Rhine Embayment (AHORNER et al., 1983; PLENEFISCH & BONJER, 1997; HINZEN, 2003).

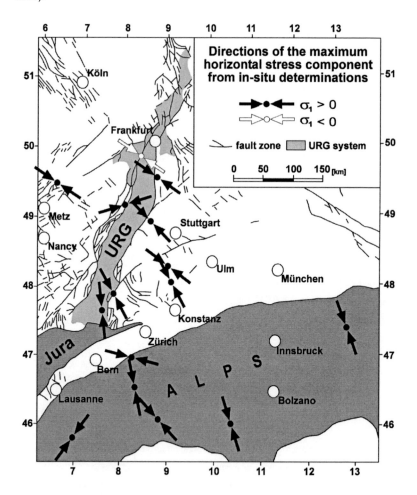

Figure 1.11: Directions of the maximum horizontal stress component derived from near surface door stopper measurements (after ILLIES & GREINER, 1979). The general NW to NNW orientation of S_H is superimposed by secondary variations along the URG and in its south (ILLIES & GREINER, 1979).

Studies of the in-situ stress state of the URG and surrounding areas based on the inversion of focal mechanisms suggest a transtensional dominantly strike-slip stress state with the maximum horizontal stress direction (S_H) oriented NNW-SSE characterising in the northern Alpine foreland (Swiss Jura and Molasse; PLENEFISCH & BONJER, 1997). The southern URG is also characterised by transtension but with a slightly reduced strike-slip component and S_H orientation NW-SE (PLENEFISCH & BONJER, 1997). The northern URG and the Rhenish Massif, where S_H is oriented NW-SE to WNW-ESE, are also believed to be in transtension, but with only a minor strike-slip component (DELOUIS et al., 1993; PLENEFISCH & BONJER, 1997; HINZEN, 2003).

Recent modelling of strain rates across the URG and Lower Rhine Embayment based on velocity measurements of continuously operating GPS stations also indicates a change from a NW–SE oriented compressional regime in the URG to an extensional regime in the Rhenish Massif and the Lower Rhine Embayment (TESAURO et al., 2005). This systematic change in kinematic behaviour along strike of the URG is generally argued to be related to the distance from the Alps. It is further argued that the relative magnitudes of S_H and S_v due to Alpine compressional forces change such that S_H decreases with respect to S_v as the distance to the Alps increases (AHORNER et al., 1983; DELOUIS et al., 1993; PLENEFISCH & BONJER, 1997).

In the central segment of the URG, at the western border, an extensive set of in-situ stress data exists from the experimental geothermal reservoir site at Soultz sous Forêts. At this site, large scale injection experiments as well as the analysis of borehole images and induced seismicity are integrated to provide a well constrained characterisation of the in-situ stress field to a depth of 5 km. In total, 4 boreholes were drilled to a final depth of up to 5 km. Because of the large diversity of the data obtained, which has been used by various groups to estimate the in-situ stress state, published principal stress directions show a large variability (e.g. RUMMEL & BAUMGÄRTNER, 1991; HEINEMANN, 1994; HELM, 1996; GAUCHER et al., 1998; Table 1.1).

Publication; used method	Well; depth interval	Direction of S_H
RUMMEL & BAUMGÄRTNER (1991); Hydraulic re-opening	GPK1; 1376-2000 m	N155° +/- 3° or N176° +/- 6°
KLEE & RUMMEL (1993); Hydraulic fractures & Hydraulic re-opening	GPK1 & EPS1; 1376-3500 m	N135° - N157.5°
JUNG (1991); Hydraulic fractures	GPK1; 1968-2000 m	N170°
MASTIN & HEINEMANN (1988); Drilling induced tensile fractures	GPK1; 1420-2000 m	N169° +/- 21°
TENZER (1991); Drilling induced tensile fractures	GPK1; 1450-2000 m	N169° +/- 11°
HEINEMANN (1994); Hydraulic fractures, drilling induced fractures	GPK1; 1420-2000 m	N170°
NAGEL (1994); Drilling induced tensile fractures	GPK1; 2000-3590 m	N181° +/- 22°
CORNET & JONES (1994); Hydraulic fractures, thermal fractures induced by large scale injection, orientation of micro-seismic cloud	GPK1; 1900-3500 m	N175°
GENTER & TENZER (1995); Drilling induced tensile fractures	GPK2; 1420-3880 m	N175° +/- 17°
HELM (1996); focal mechanisms inversion	none; 2300-3500 m	N125° +/- 20°
GAUCHER et al. (1998); Shear wave splitting	none; 1350-1450 m	N180°
BERARD & CORNET (2003); Thermal borehole cross-section elongation	GPK2; 1422-2700 m	N164° +/- 18°
BERARD & CORNET (2003); Compressive borehole breakouts	GPK1 & GPK2; 3020-3650 m	N185° +/- 7° (GPK1) N185° +/- 25° (GPK2)
VALLEY & EVANS (2007); Breakouts and drilling induced tensile fractures	GPK3 & GPK4 1432 – 5248 m	N169° +/-14 (average of GPK3 & GPK 4)

Table 1.1: Summary of the maximum horizontal stress orientation determinations obtained at Soultz sous Forêts (modified after CORNET et al., 2007).

Principal stress direction determinations range from N125°E to N185°E and also several different magnitude estimations have been proposed (Cornet et al., 2007). For example, Rummel & Baumgärtner (1991) proposed the following relationship of the stress components with depth, valid for depths greater than 1458 m:

$S_{v\,[MPa]}$ = $\quad$ 0.024 $z_{[m]}$

$S_{H\,[MPa]}$ = $\quad$ 24.8 + 0.0198 ($z_{[m]}$ - 1458)

$S_{h\,[MPa]}$ = $\quad$ 15.1 + 0.0179 ($z_{[m]}$ - 1458)

In contrast, Heinemann (1994) proposed the following relationship of the stress components with depth:

$S_{v\,[MPa]}$ = $\quad$ 33.8 + 0.0149 ($z_{[m]}$ - 1377)

$S_{H\,[MPa]}$ = $\quad$ 23.7 + 0.0336 ($z_{[m]}$ - 1458)

$S_{h\,[MPa]}$ = $\quad$ 15.8 + 0.0255 ($z_{[m]}$ - 1458)

A compilation of in-situ stress data derived from various sampling depths for the URG area from different sources, such as borehole breakouts, hydro fracturing, overcoring and focal mechanisms, has been released by the World Stress Map (Reinecker et al., 2004; Figure 1.12). This data shows a general NW to NNW orientation of S_H and indicates that S_H rotates ~ 15° anticlockwise from the southern to the northern URG (Müller et al., 1992).

The studies reviewed argue that the URG is not characterised by a discrete stress field and has only minor influence on the orientation of present-day regional stress field (e.g. Larroque et al., 1987; Plenefisch & Bonjer, 1997). Most models of the recent URG kinematics suggest that it is acting as a sinistral shear zone in a strike-slip regime (Illies, 1974a; Illies & Greiner, 1976, Schumacher, 2002) that is driven by the present-day regional stress field, which is mainly controlled by plate boundary loads.

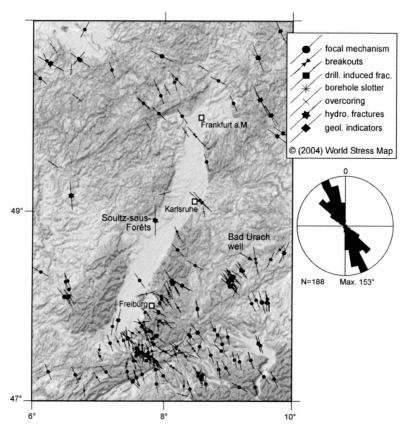

Figure 1.12: In-situ stress indicators compiled by the WSM (REINECKER et al., 2004). In general, a NW–SE to NNW-SSE orientation of the maximum horizontal stress component (S_H) is visible for the URG area.

1.4.3 Seismicity of the Upper Rhine Graben

The URG is characterised by low to moderate intra-plate earthquake activity (AHONER, 1983; BONJER, 1997; LEYDEKER, 2005). The historical seismic catalogue of the URG area enfolds a period since 800 AD (e.g. LEYDECKER, 2005A). The instrumental and historical records of seismic events show a wide distribution of earthquakes occurring over the entire region (Figure 1.13).

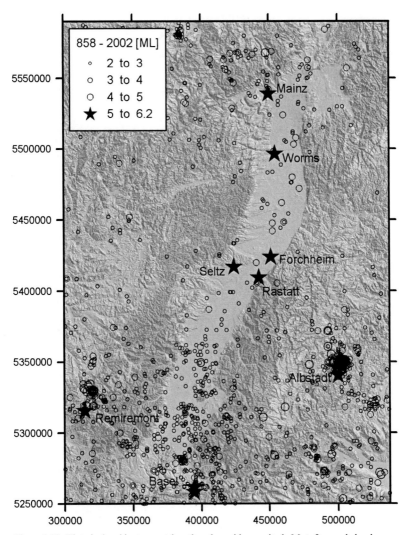

Figure 1.13: Historical and instrumental earthquakes with magnitude $M_L > 2$ recorded or known in the URG area between 858 and 2002. For conversion of I_o to M_L the formula $M_L = 0.636 * I_o + 0.4$ of RUDLOFF & LEYDECKER (2002) has been used. Data source: FRACASSI et al. (2005) and LEYDECKER (2005A).

Along the strike of the URG the seismicity pattern changes and shows regional variations. In the southern part of the graben, the seismicity is slightly elevated as compared to the northern part and is characterised by a wide distribution of small earthquakes (M_L 2 to 4). These earthquakes commonly occur between 5 and 15 km depth but may locally occur in greater depths (see section 1.3.3; BONJER et al., 1984; PLENEFISCH & BONJER, 1997). In the central graben segment a few scattered earthquakes with focal depths of 10 to 15 km are documented (BONJER et al., 1984). In this segment several damaging earthquakes have occurred in historical times (e.g. Rastatt 1933, Seltz, 1952; Figure 1.14).

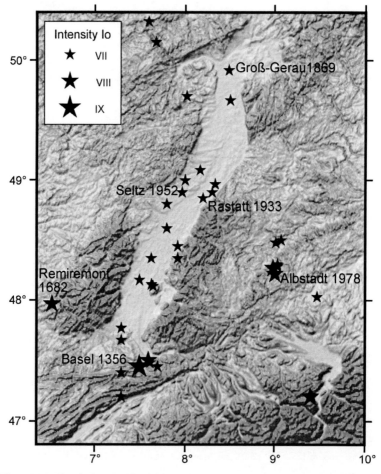

Figure 1.14: Historical earthquakes with intensities I_o > VII and documented damages from 1021 – 1992 AD in the URG area (after FRACASSI ET AL, 2005; LEYDECKER, 2005A; SCHWARZ et al., 2006). Note the concentration of damaging earthquakes in the southern and central segments of the URG, whereas damaging earthquakes were less frequent in the northern segment.

The northern part of the URG is also characterised by small earthquakes (M_L < 4) but their occurrence is less frequent as compared to the southern part. A maximum focal depth of 20 km is reported in the northern segment with the majority of the seismicity occurring in the upper 10 km (BAIER & WERNING, 1983; BONJER et al., 1984). In the northern segment of the URG the seismicity is mainly restricted to the graben with significantly lower seismicity in the shoulder. In the vicinity of the URG, areas with the highest documented seismicity are situated at its southern termination, near the city of Basel, to the west of the Vosges Mountains and in the Swabian Alp; where earthquakes with magnitudes M_L > 5 or intensities I_0 > VII occurred frequently in historic times (e.g. LEYDECKER, 2005A; Figure 1.14). The strongest earthquake documented in the vicinity of the URG occurred in 1356 at its southern termination near the city of Basel. Several studies suggest a potential magnitude of M > 6 for this event (e.g. MEGHRAOUI et al., 2001). The strongest historical earthquake documented within the URG occurred in 1933 near the city of Rastatt with an intensity I_0 of VII and an estimated local magnitude M_L of 5.4 (LEYDECKER, 2005A; Figure 1.14). However, large earthquakes in the URG area are rare and rather an exception. Several authors have proposed that creep is the dominant mechanism in this area (e.g. AHORNER, 1975; BONJER et al., 1984; FRACASSI et al., 2005; PETERS, 2007). They suggest that due to the high fracturing of the graben sufficient energy for large earthquakes cannot be accumulated but releases through smaller earthquakes (M < 5) or creeping motions.

1.5 Thesis outline

This Ph.D. thesis focuses on the modelling of the contemporary stress state of an abandoned continental rift structure, the Upper Rhine Graben, and its present-day reactivation and kinematic behaviour. In this context, modelling of the reactivation potential of pre-existing faults systems associated with the graben structure is one of the main goals. Furthermore, based on the evaluation of the fault reactivation potential, a possible contribution of 3 dimensional mechanical earth modelling to earthquake hazard assessment is investigated.

To construct and solve 3 dimensional mechanical earth model (MEM) many different data sets such as the geometric structure, material properties and boundary conditions need to be integrated. In general, such a data base is required to:

- enable the construction of a geological or geometrical model volume
- provide constraints on model boundaries

- provide calibration data
- provide independent data for model evaluation (benchmarking).

Another task of this Ph.D. thesis is the development of a method and work process for the construction of complex model geometries based on the different data types available. In order to establish a procedure that is independent of local computing and software facilities, the work-flow used in this thesis is predominantly based on commercial software packages. The integration of the geometrical data and the construction of the geological models are conducted using the geological modelling package gOcad®. For the finite element discretisation of the geological model obtained from gOcad, the pre- and post-processing software HyperMesh® was used. To solve the 3 dimensional geomechanical models, the finite element code ABAQUS™ was used. The modelling results were visualized and post-processed using the in-house post processing package GeoMoVie© in order to enable the analysis and interpretation of the models conducted.

In chapter 2, "Tectonic stresses and fault reactivation", a brief introduction is given on crustal stresses, their definition, determination and classification. The main focus of this chapter is on the theoretical background of fault reactivation. Two approaches of shear failure reactivation evaluation, independent of the rheological parameter of fault surfaces, are discussed.

In chapter 3, "Finite element modelling of the intra-plate stress state", a summary of the finite element method is given. The first part of chapter 3 refers to the basic principles and the use of this method in Earth science. Since the work presented in this Ph.D. thesis uses a commercial finite element code (ABAQUS™) the advantages and disadvantages of this software package are demonstrated. This includes the influence of mesh quality and the implementation of contact problems as well as the ABAQUS™ implementation of the material models used (elasticity and elasto-plasticity). The second part of chapter 3 refers to the approach of multi-scale modelling, nesting or sub-modelling using ABAQUS™. The consequences of this approach on the boundary conditions and the model geometries are discussed. The third part of this chapter examines boundary conditions, model calibration, model benchmarking and the initial model conditions.

In chapter 4, "Simulating the in-situ state of stress of the Upper Rhine Graben (regional model)", the present-day tectonics of the URG and kinematic behaviour of first order fault structures are investigated by conducting a 3D finite element analysis (FEA). The datasets used to construct the regional model as well as the geological setting constraining the internal model set-up and boundary conditions are presented and discussed. The first part of this chapter focuses on the model geometry implemented and the modelling parameters and boundary conditions used. In the second part, the 3D finite element model of the entire URG is evaluated against independent data such as relative surface uplift. Furthermore, the distribution of localised deformation predicted is compared to observed faulting and earthquake activity. The results of the regional model study reveal a significant influence of the various topographies (surface, Conrad and Moho) and the geometric structures implemented on the present-day stress state and the recent kinematic behaviour of the URG. Furthermore, the boundary condition analysis of the URG model shows that the magnitude of Moho uplift is a critical loading condition on the regional kinematic model.

Chapter 5, "Predicting the in-situ stress state and the fault reactivation potential in the central segment of the URG (*sub-model A*)", addresses the contribution of second order structures to the kinematic behaviour of the central URG and the fault reactivation potential in this region. A local scale, high resolution model of the central URG, including complex fault geometries and several geological bodies with different rheological properties, is used to evaluate local stress perturbations. To transfer the state of in-situ stress, the boundary conditions of this local scale model is inferred from the regional scale URG model. The results of the local scale model are evaluated against independent data and compared to the regional scale modelling results. Furthermore, earthquake activity in the central part of the URG is investigated using the concept of static Coulomb Failure Stress change. The modelling results indicate that within the present-day stress field, the central URG segment is more compressional than the overall graben trend. Therefore, highly variable faulting mechanisms including extension, strike-slip and inversion of pre-existing extensional faults are predicted. The comparison of seismogenic faults in the area with the results of a slip tendency analysis, suggest that damaging earthquakes in this region are possibly associated with not favourably oriented fault segments.

Chapter 6, "Predicting the in-situ stress state and the fault reactivation potential in the northern segment of the URG (*sub-model B*)", addresses a more complex "test area" for the multi-scale modelling approach. This model covers the northern URG segment that consists of several fault-blocks and sub-basins. Based on the study of PETERS (2007), a detailed dataset of the neotectonic and contemporary tectonic behaviour for this model area is available. Considering this dataset, relative block motions predicted by the model can be evaluated in great detail. Furthermore, the effect of a possible magnitude 6 palaeo-earthquake proposed by PETERS et al. (2005) is investigated using the method of static Coulomb Failure Stress change. The modelling results indicate that under the present-day stress field, the northern URG segment forms a releasing bend. Here, the lateral compression on the graben is reduced kinematically due to a change in strike of the graben with respect to the central segment. In the northern graben segment the Eastern Border Fault (EBF), which is predicted to be most active, coincides with the location of the most pronounced Quaternary depocentre. Along the EBF increased uplift is predicted for the associated graben shoulder corresponding to the Odenwald basement high due to the oblique orientation of the graben axis with respect to the axis of highest Moho uplift. Furthermore, high slip tendency and dilation tendency values for this segment of the EBF indicate increased tectonic activity.

In chapter 7, "Integrated discussion and conclusion of the modelling results", the results of all three models (chapters 4, 5 and 6) are synthesised and evaluated. In the first part of this chapter, the objectives of this thesis and the applied modelling approach are summarised. In the second part of the chapter, the results of the individual models are presented and discussed and kinematic models for the individual model-scales are developed. Additionally, the results of the two pseudo dynamic earthquake models conducted are integrated and a possible contribution of finite element modelling to earthquake hazard assessment is discussed. Finally, the main conclusions are summarised and possible improvements of the modelling approach applied are discussed.

TECTONIC STRESSES AND FAULT REACTIVATION

This chapter focuses on the definition of stress, the theoretical concepts for the classification of tectonic stresses in the Earth's crust and on the theoretical background of the evaluation of fault reactivation potential. Following a short introduction on stresses in the Earth's crust and their determination, focus of this chapter are:

- The definition and description of stresses.
- The classification of tectonic stresses combining the approaches of ANDERSON (1905) and BOTT (1959).
- A general introduction to fault reactivation.
- The evaluation of fault reactivation potential of pre-existing fault surfaces using the slip tendency approach by MORRIS et al. (1996).
- The dynamic approach of fault interaction (ΔCFS) by REASENBERG & SIMPSON (1992).

2.1   Crustal stresses

Deformation of the Earth's crust is the result of the action of two types of applied forces. Body forces, such as gravity, act throughout the volume of the considered body or continuum and thus their magnitude is proportional to the volume or mass of the body. They arise as a consequence of the continuum being placed in a force field. In contrast, surface or contact forces act on arbitrary surfaces that bound a considered volume. These external and internal forces, acting on a considered volume of rock within the Earth's crust, induce stresses. They are the primary source of deformation and failure within the Earth's crust and introduce contact forces on pre-existing heterogeneities such as fault surfaces or other zones of weakness.

In addition to physical boundary conditions, the stresses are dependant on the mechanical properties of the rock volume. Since the mechanical properties of the crust are vertically and horizontally non-uniform, the state of stress, the stress magnitudes and orientation exhibit great spatial variability. Furthermore, the spatial variability of external and internal boundary conditions such as temperature or pressure, contribute to the observed heterogeneous stress pattern.

The determination of the state of stress in the Earth's crust is limited to the observation of strain, as stresses cannot be directly measured or observed. By linking stress through material properties with the observation of strain, stresses can be inferred.

Using this approach, the orientations of the principal stress axes (σ_1, σ_2 and σ_3) and their relative magnitudes can be determined. The information of the spatial orientation of the principal stress axes and the stress ratio R, derived from their relative magnitudes, can be used to constrain the tectonic regime in which the observed strain occurs (see section 2.1.2). Since strain is observed locally, information of the stress state can only be obtained on a local scale and may be related to local heterogeneities of the crust. The following different approaches to determine a local stress state can be applied. The first arrival of seismic waves can be used to determine earthquake focal mechanism solutions. A focal mechanism solution is the 3 dimensional spherical representation of the polarity distribution of the first arrival of P-waves induced by an earthquake. Using a grid of seismometers, the compressive and dilative quadrants of the sphere and the principal strain axes (P, B and T) can be identified. Inverting seismic wave forms of a broad frequency band for the moment tensor, the focal mechanism can be determined using even a small number of seismic recordings (BARTH, 2006). Under the assumption that the observed earthquake occurred on an optimally oriented plane one could link the principal stress axes directly to the principal strain axes identified. Using the technique of stress inversion, based on several earthquake focal mechanisms, the stress state (spatial orientations of the principal stress axes and R-ratio) of a rock volume can be inferred (McKENZIE, 1969). The disadvantage of this technique is that it assumes all considered strain events to have occurred in a homogeneous medium under equivalent stress conditions. Besides seismological methods of stress determination, in-situ stress measurements or methods linked to drilling techniques can be applied. In-situ stress measurements such as over-coring determine the strain relaxation after a probe of rock is separated from the surrounding material. In-situ stress orientations from greater depth can also be determined by the analysis of drilling induced wellbore failures, such as borehole breakouts and drilling induced tensile wall fractures. In combination with the technique of hydraulic fracturing, which enables the determination of the least principal stress magnitude, the stress state along a wellbore path can be investigated (e.g. ZOBACK & HEALY, 1992; BRUDY et al., 1997). The disadvantage of these methods is the limitation of applicability to the upper part of the crust.

Using the approach of finite element modelling to simulate the state of stress within a geological volume, it is essential to calibrate the model with in-situ stress data. The kinematic behaviour of fault surfaces investigated is dependant on the stress field, which induces the contact forces on that particular surface. Therefore, in-situ stress determinations provide constraints for failure and reactivation prediction using 3 dimensional mechanical earth models (MEM).

2.1.1 Sources of tectonic stresses

Within the Earth relative movements of rock units occur along systems of discrete fault surfaces or in zones of distributed deformation. The sources of this relative movement are founded in large-scale thermal instabilities and convective flow within the Earth's mantle (e.g. EISBACHER, 1996). These thermally driven processes transfer tectonic stresses interdependently on the crust. Gravity acting on a volume of rock is a primary source for introducing stresses (e.g. RANALLI, 1992; GÖLKE & COBLENTZ, 1996). On a global scale, lateral density and thickness variations of the lithosphere are compensated by isostasy. However, at a crustal scale, these lateral variations lead to a potential energy difference inducing a lateral pressure gradient, which is compensated by tectonic stresses. Examples for gravity-induced forces acting on a plate boundary are the compressive ridge-push forces, which are one of the important forces driving the plates (TURCOTTE & SCHUBERT, 2002). Differential gravitational loading can also introduce forces along passive continental margins (LISTER, 1975; BOTT, 1991). GÖLKE & COBLENTZ (1996) identified the stresses induced by differential gravitational loading to be the main contribution to the contemporary stress field affecting NW Europe. In the subducting oceanic lithosphere a density contrast can induce tensional slab-pull forces. Another important contribution to a plate wide (regional) stress field is the so-called collisional resistance, which is associated with convergent plate boundaries. Excess compressive stresses induced by the collisional resistance, if not compensated by deformation in the orogen, can be transferred deep into the colliding plates. For NW Europe ZIEGLER (1987) demonstrated, based on the analysis of inverted sediment basins, that these stresses can be effective more than 1000 km away from the active margin. GÖLKE & COBLENTZ (1996) identified this type of stress to have a secondary influence of the present-day stress field of NW Europe. The influence of other sources of tectonic stresses is more localised. During the process of subduction, the tensional trench suction force is induced in the overriding plate and initiates the evolution of back-arc basins. Surface processes such as erosion, sedimentation and glaciation can generate differential loading on the Earth's surface. These loads are the source for another type of localised tectonic stresses, the flexural stresses (BODINE et al.,1981; CLOETINGH et al., 1982). Another source of flexural stresses are the bending stresses associated with the buckling of the lithosphere during the process of lithospheric folding (CLOETINGH et al., 1999).

2.2 Definition and description of stress

Stress, σ, is defined as the resistance against surface forces (*SI* unit Pascal, 1Pa = 1Nm^{-2}). The algebraic sign convention for stresses in engineering literature is tensile stresses being positive and compressive stresses being negative. However, in the Earth's crust the most common stresses are compressive. Therefore, in geosciences a different algebraic sign convention with compressive stresses being positive is used (JAEGER & COOK, 1979). This sign convention is followed in this study.

Stress can be expressed by the traction vector $\vec{T}_{(n)}$ (also termed stress vector $\vec{S}_{(n)}$) which is defined by the limit:

Equation 2.1
$$\vec{T}_{(n)} = \lim_{\Delta A \to 0} \frac{\Delta \vec{F}}{\Delta A} .$$

Thus the traction vector $\vec{T}_{(n)}$ is the force per unit area acting on a plane that is oriented such that its outward directed normal is n. The traction vector $\vec{T}_{(n)}$ can comprise any angle to an arbitrary surface. Therefore, $\vec{T}_{(n)}$ can be partitioned into the normal stress component, acting perpendicular to the surface and the shear stress component acting parallel to the surface. Cauchy demonstrated that the traction vector on any arbitrary surface can be determined by

Equation 2.2
$$\vec{T}_{(n)} = \sigma_{ij}\hat{n}_j .$$

Where $\hat{n}_j$ represents the unit normal vector of any arbitrary surface and σ_{ij} describes the three dimensional stress tensor at any point as:

Equation 2.3
$$\sigma_{ij} = \begin{pmatrix} \sigma_{xx} & \sigma_{xy} & \sigma_{xz} \\ \sigma_{yx} & \sigma_{yy} & \sigma_{yz} \\ \sigma_{zx} & \sigma_{zy} & \sigma_{zz} \end{pmatrix} .$$

This is Cauchy's fundamental stress theorem (e.g. DAVIS & SELVADURAI, 1996), such that the matrix σ_{ij} describes the 3 dimensional stress tensor at any given point. Furthermore, Cauchy demonstrated that the stress tensor is symmetric and thus only six tensor components are required to define the state of stress at any point.

2.2.1 Principal stresses

By applying a transformation of the principal axes of the stress tensor given in equation 2.3, the shear stresses in the coordinate planes vanish and the stress tensor consists only of the normal stresses σ_1, σ_2 and σ_3, which are referred to as the principal stresses:

Equation 2.4

$$\sigma^P = \begin{pmatrix} \sigma_1 & 0 & 0 \\ 0 & \sigma_2 & 0 \\ 0 & 0 & \sigma_3 \end{pmatrix}$$

The three principal stresses are defined as maximum principal stress σ_1, intermediate principal stress σ_2 and minimum principal stress σ_3.

2.2.2 Mean stress

The mean stress σ_m is the first stress invariant, and is sometime termed the pressure P, although sensu-stricto this corresponds to the hydrostatic pressure under the condition $\sigma_1 = \sigma_2 = \sigma_3$. The mean stress causes volume changes in rocks and is defined as the arithmetic mean of the three principal stresses:

Equation 2.5

$$\sigma_m = P = \frac{\sigma_1 + \sigma_2 + \sigma_3}{3},$$

and is equivalent to the trace of the stress matrix.

2.2.3 Von Mises Stress

The von Mises Stress σ_M is the second invariant of the stress tensor and is often used to predict plastic failure (e.g. RANALLI, 1995). Von Mises Stress is a scalar value derived from the principal stress components as:

Equation 2.6

$$\sigma_M = \sqrt{\frac{(\sigma_1 - \sigma_2)^2 + (\sigma_2 - \sigma_3)^2 + (\sigma_3 - \sigma_1)^2}{2}}.$$

2.2.4 Differential stress

The differential stress σ_d is the main factor responsible in producing shear stresses, which causes fracturing in materials. The differential stress is the difference between the maximum principal stress and the minimum principal stress:

Equation 2.7 $\qquad\qquad\qquad\qquad\sigma_d = \sigma_1 - \sigma_3$.

2.2.5 The Mohr Circle of Stress

The Mohr Circle of stress is a graphical approach to derive the quantitative relationship between normal stress (σ_n) and shear stress (τ) acting on a point or an arbitrarily oriented plane when the principal stress magnitudes are known. Mohr demonstrated that the state of stress can be represented in a 2 dimensional coordinate system with the x-axis representing the normal stress (σ_n) and y-axis representing the shear stress (τ; Figure 2.1a).

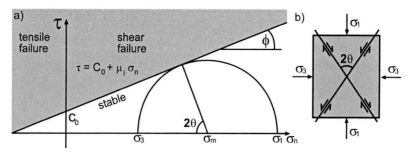

Figure 2.1: a) The Mohr Circle of Stress as graphical representation of the relationship between shear and normal stress. The Mohr Circle construction is used to demonstrate shear failure to occur when the Coulomb failure criterion is reached. b) Two conjugated shear planes are possible to form at an angle of $\pm\theta - 45° - \phi/2$ after the shear strength of a rock sample is reached.

The Mohr Circle also demonstrates that maximum shear stress (τ_{max}) occurs on two perpendicular planes ($2\theta = 90°$), oriented $\pm45°$ to the principal stress axis (Figure 2.1b). The magnitude of the maximum shear stress is only dependent on the differential stress:

Equation 2.8 $\qquad\qquad\qquad\qquad \tau_{max} = \dfrac{\sigma_1 - \sigma_3}{2}$

A Mohr Circle analysis can be applied to many problems of geomechanical or geodynamic significance such as fracturing and faulting processes or fault reactivation. When the differential stresses cannot be accommodated by elastic deformation, plastic deformation, i.e. brittle failure will occur. Two types of fracturing can be addressed in a Mohr Circle analysis using various failure criteria: tensile (Mode I) failure and shear

(Modes II and III) failure (Figure 2.2).

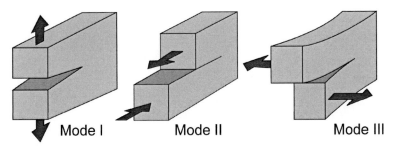

Mode I   Mode II   Mode III

Figure 2.2: The three basic types of fracture tip deformation, Mode I (tension, opening), Mode II (in plane shear, sliding) and Mode III (out of plain shear, tearing).

Tensile fractures form normal to the minimum principal stress when the tensile strength (T) of the material is exceeded. Coulomb proposed shear fractures to form dependent on the normal stress when the shear stress in the rock volume exceeds the cohesive strength (C_0) and the internal frictional resistance (μ_i; i.e. Coulomb failure criterion):

Equation 2.9 $$\tau = C_0 + \mu_i \sigma_n.$$

The internal frictional resistance (μ_i) can be derived from the angle of internal friction (ϕ_i) by

Equation 2.10 $$\mu_i = \tan\phi_i.$$

Considering the geometry of the Mohr Circle (Figure 2.1a), the relationship between the angle of internal friction and the failure angle is given by

Equation 2.11 $$\theta = 45° - \frac{\phi_i}{2},$$

whereas the generation of two conjugated shear planes is possible at angles of $\pm\theta$ relative to σ_1 (Figure 2.1b).

In order to distinguish between tensile and shear fracture generation a combined Griffith-Coulomb failure criterion can be used. Griffith (see JAEGER & COOK, 1979) demonstrated that the differential stress is the most important factor for the generation of the different types of fractures (Figure 2.3). Tensile failure can only occur when the differential stress is less than four times the tensile strength (T), whereas shear failure can only occur when the differential stress is larger than four times the tensile strength (T; see section 2.1.4).

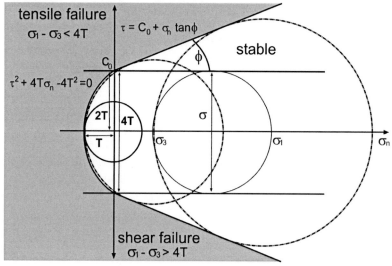

Figure 2.3: Based on the acting differential stress and the assumed tensile strength, the combined Griffith-Coulomb failure criterion can be used to distinguish between the generation of different fracture types. Tensile failure occurs only when $\sigma_d < 4T$ (bold circle), whereas shear failure occurs when $\sigma_d > 4T$ (dashed circles).

2.2.6 The fracture potential method (FP)

The type and spatial distribution of fractures generated in a brittle tectonic system is critical for a complete understanding of the systems' behaviour. The concept of fracture potential (FP; CONNOLLY, 1996; ECKERT & CONNOLLY, 2004), is based on the combined Griffith/Navier-Coulomb failure criterion (JAEGER & COOK, 1979; PRICE & COSGROVE, 1990), and enables distinction between different types of fractures (shear or tensile). Since in many FE modelling studies only linear elastic rheologies are used, the FP can be used to investigate the occurrence of brittle failure without implementing plastic material models or second order fractures. The method of FP quantifies the likelihood of fracture generation. Brittle deformation occurs in the tensile regime when the differential stress, σ_d ($\sigma_1 - \sigma_3$), is less than four times the tensile strength T (Figure

2.4a). The tensile FP (tFP) magnitude is then defined as the ratio of σ_3 to the tensile strength T of the material:

Equation 2.12
$$tFP = \frac{\sigma_3}{T} \, .$$

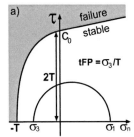
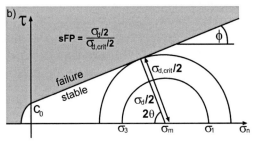

Figure 2.4: Mohr diagrams illustrating the concept of fracture potential (ECKERT & CONNOLLY 2004). a) In the tensile regime ($\sigma_d < 4T$) tFP is the ratio between σ_3 and T and is negative by convention. At tensile failure $\sigma_3 = -T$ and tFP = -1. When $\sigma_3 > 0$ tensile failure is not possible and tFP = 0. b) In the shear regime ($\sigma_d > 4T$) sFP is the ratio between the critical differential stress at failure σ_{dcrit} and the observed differential stress σ_d and is positive by convention. When $\sigma_d < 4T$ shear failure is not possible and sFP = 0.

In the shear regime (Figure 2.4b), where σ_d is greater than four times the tensile strength, the shear FP (sFP) magnitude is defined by the relationship between the critical differential stress at failure σ_{dcrit} and the modelled differential stress σ_d:

Equation 2.13
$$sFP = \frac{\sigma_d/2}{\sigma_{dcrit}/2} = \frac{\sigma_d}{2(c_0 \cos\phi + \sigma_m \sin\phi)} \, .$$

The FP is calculated for the state of stress at each data point of the model. The FE model results contain the stress and strain tensor at each integration point. The differential stress, mean stress and the stress orientation can be derived directly from the modelled stress tensor. During post-processing, idealised elastic rheologies, including cohesion (c_0, assumed to be twice the tensile strength; PRICE, 1966; PRICE & COSGROVE, 1990) and a coefficient of internal friction (μ_i), are used to calculate the FP:

Equation 2.14
$$C_0 = 2T \, .$$

2.2.7 Classification of tectonic stresses: Regime Stress Ratio (RSR)

The state of stress in an elastic medium can be described by 3 orthogonal compressive or tensional vectors, the principal stress vectors (σ_1, σ_2 and σ_3). ANDERSON (1905) argued that in general the spatial orientation of these principal stress vectors is a function of the Earth's free surface, which can carry no shear stresses. Thus, the maximum horizontal (S_H), the minimum horizontal (S_h) and the vertical stress components (S_V) are suggested to be parallel to the principal stress vectors. Based on this assumption, ANDERSON (1905) established the traditional classification of stresses within the Earth's crust by their orientation towards the earth's free surface and defined the 3 possible tectonic regimes (extensional, strike-slip and thrusting) with their characteristic deformation pattern (Figure 2.5).

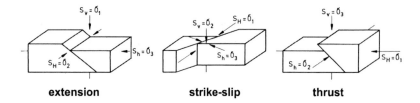

extension   **strike-slip**   **thrust**

Figure 2.5: The 3 Andersonian stress regimes and the associated faulting types. Close to the surface, which can carry no shear stresses, the principal stress vectors are parallel to the vertical and horizontal stress components. Figure modified after EISBACHER (1996).

To use the spatial information of the orientation of the principal stress vectors for further calculations in this study, values of 0, 1 and 2 are assigned for extensional, strike slip and thrusting regime respectively.

The state of stress in the earth's crust can also be classified by the so-called shape ratio of the stress ellipsoid. This stress ratio (R), defined after BOTT (1959) as

Equation 2.15

$$R = \frac{(\sigma_2 - \sigma_3)}{(\sigma_1 - \sigma_3)}$$

and describes the ratio of the principal stress magnitudes, where the R-value ranges between 0 and 1. The R-value describes the purity of the Andersonian stress regimes, with the R-value equal to 0.5 under ideal extensional, strike-slip and thrusting conditions respectively. For the strike-slip regime an R-value smaller or greater than

0.5 is indicating the transition towards the two other stress regimes. An R-value close to the possible minimum or maximum (0 or 1) indicates an unstable state of stress with 2 of the principal stress vectors having the same magnitude (Figure 2.6).

Considering the results of a 3D finite element model, the stress tensor is known at each integration point. To visualise the three dimensional stress state of the model, the Regime Stress Ratio value (RSR) is used. The RSR value, calculated at each data point in the model, is a combination of the Andersonian stress regime and shape ratio of the stress ellipsoid. The RSR value is then found by combining the stress regime and the stress ratio through the following relationship (e.g. SIMPSON, 1997):

Equation 2.16 $$RSR = (regime + 0.5) + (-1)^{regime}(R - 0.5).$$

The resulting RSR value is continuous in the range 0 to 3 and describes the type of deformation expected from radial extension to constriction (Figure 2.6). The calculation of RSR values enables the determination of the type of faulting for each structure and region within the model. The advantage of using this value for the interpretation of relative stress magnitudes is that it enables better resolution of the expected faulting type, particularly in regions where transpression and transtension are important (Figure 2.6).

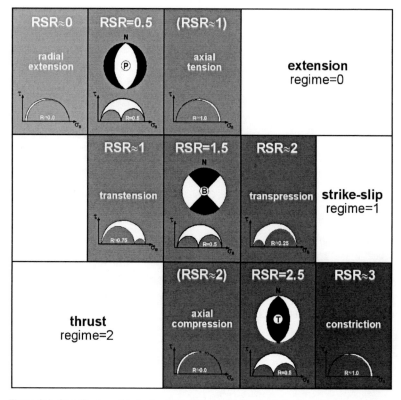

Figure 2.6: Classification of tectonic stresses. Rows display the 3 Andersonian stress regimes (extension, strike-slip and thrust), defined by the spatial orientation of the principal stress axes. Additionally, the state of stress is defined by the stress ratio (R-value). By combining both approaches of classification through equation 2.16, a parameter (RSR) can be defined which describes the 3 dimensional state of stress dependent on orientation and relative magnitudes of the principal stress axes.

2.3 Reactivation of faults

Crustal stresses of an arbitrary geological source introduce contact forces on a pre-existing fault surface or zone of weakness. Possible failure along these pre-existing discontinuities is dependent on the stress state, physical properties of the fault surface, the pore fluid pressure and the spatial orientation of the fault surface. If the induced contact forces overcome the strengths of a fault zone, slip or dilation can occur.

Detailed understanding and development of adequate physical models of the possible kinematic behaviour of pre-existing fault surfaces is of great economical and societal importance. In many regions, seismic hazard assessment is a key societal task. Here, an adequate estimation of the fault reactivation potential can contribute amongst other methods to the assessment of the possible regional hazard (MORRIS et al., 1996; FERRILL et al., 1999). The evaluation of fault reactivation potential can help to constrain the maximum possible earthquake expected (LEYDECKER, 2005B). In addition, earthquake interactions and seismic sequences can be investigated. Fault reactivation prediction is also applicable to multiple tasks in reservoir geology, where for example the effect of faulting on the reservoir geometry is important to assess. Furthermore, changes in fault stability due to stimulation or production processes can be predicted.

To address the problem of fault reactivation in detail, well constrained input parameters such as material properties, physical properties of the pre-existing fault surfaces, pore fluid pressure and the geometrical structure of the study volume are necessary. Furthermore, it is necessary to quantify the state of stress introducing the contact forces along the fault surface.

2.3.1 Theory of fault reactivation

Reactivation of a pre-existing fault surface is dependent on the orientation and magnitude of the acting stress field, the fault geometry and physical properties of the fault surface such as the coefficient of friction and the cohesion. Additionally, pore fluid pressure can reduce the shear strength of a fault surface. The coefficient of friction μ is defined to be the ratio of shear stress to normal stress:

Equation 2.17
$$\mu = \frac{\tau}{\sigma_n} .$$

τ magnitude of shear stress

σ_n magnitude of normal stress

The coefficient of friction μ is required to:
- initiate a sliding surface (internal friction; μ_i),
- initiate sliding on the surface (static friction; μ_{static}) and
- maintain sliding on the surface (dynamic friction; $\mu_{dynamic}$).

Currently, no friction law based on micromechanical processes is available due to the complex interactions of surface contacts, which evolve during sliding. Therefore, friction laws are only phenomenological descriptions of the friction behaviour and are based on empirical studies. Values for the coefficient of static friction for cohesionless faults are provided by empirical data of friction experiments. Values of 0.85 for μ_{static} account for pressure conditions with $\sigma_n < 200$ MPa and values of 0.6 for $\sigma_n > 200$ MPa, but can be lower if fault surfaces contain fault gouges with clay minerals (Byerlee Law; BYERLEE, 1978). Regional values for the coefficient of static friction can be calculated from inversion of earthquake focal mechanism data. Friction values determined from inversion are usually more scattered and commonly lower than 0.6 (e.g. values of 0.3 to 0.6 from earthquake focal mechanisms in the southern URG, PLENEFISCH & BONJER, 1997). However, Byerlee's Law remains a useful approximation for estimating the strength of fault surfaces for several reasons. This law is independent of lithology, strain rate and surface roughness and is valid for a wide rage of hardness and ductility of various rock types (summarized in SCHOLZ, 2002, p. 67, see references therein). Additionally, friction values determined in boreholes (deeper than 1 km) in a variety of tectonic regimes are similar to those observed in laboratory experiments (McGARR & GAY, 1978; ZOBACK & HEALY, 1984; ZOBACK AND HEALY, 1992; BRUDY et al., 1997).

To describe possible failure of a pre-existing fault surface several approaches can be used. Figure 2.7 illustrates so-called geomechanical risking parameters (HILLIS & NELSON, 2005 and references therein) which can be used to evaluate the potential of shear and tensile failure of a rock mass or pre-existing fault surface.

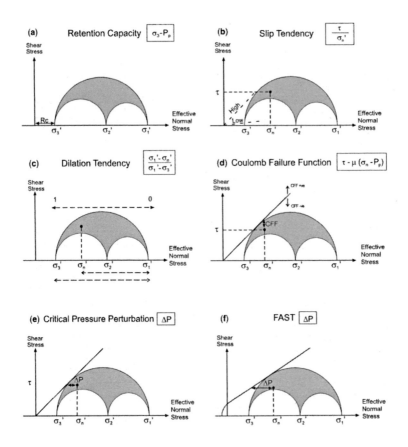

Figure 2.7: Geomechanical risking parameters after HILLIS & NELSON (2005). In this study the approaches of slip tendency (b), dilation tendency (c) and ΔCFS, based on the Coulomb failure stress / Coulomb failure function (CFF; d), are used to investigate possible shear failure. The static description of loading based on Amontons' Laws from 1699 (based on observations by Leonardo da Vinci):

Equation 2.18 $\qquad\qquad \tau = \mu_{static} \sigma_n$

describes frictional sliding on a pre-existing fault surface and relates shear to normal stress. Based on this concept MORRIS et al. (1996) introduced the slip tendency parameter (T_s; section 2.2.2), which is a quantitative measure of the likelihood of fault reactivation:

Equation 2.19

$$T_s = \frac{\tau}{\sigma_n} .$$

A dynamic approach based on the Coulomb Failure Criterion, first introduced by Coulomb in 1773 (JAEGER & COOK; 1979):

Equation 2.20

$$\tau = C + \mu_i \sigma_n$$

C   magnitude of cohesion of intact material

μ_i coefficient of internal friction

states that failure takes place when the shear stress exceeds the normal stress and the cohesion of intact rock at a specific plane. A useful approach can be deduced by considering whether a change in the stress field causes pre-existing fault surfaces to reach a stress state closer to failure. In this study the change of the stress field is considered to be caused by a nearby earthquake event. Based on this concept REASENBERG & SIMPSON (1992) developed the definition of change in Coulomb Failure Stress (ΔCFS; see section 2.2.3), which can be used as a qualitative measure of the likelihood of dynamic fault reactivation:

Equation 2.21

$$\Delta CFS = \Delta\tau - \mu\Delta\sigma_n .$$

$\Delta\tau$ change in shear stress magnitude

$\Delta\sigma_n$ change in normal stress magnitude

In order to evaluate the relative likelihood for a fault structure or extensional fracture to dilate, the so-called dilation tendency (DT) can be used (FERRILL et al., 1999):

Equation 2.22

$$DT = \frac{\sigma_1 - \sigma_n}{\sigma_1 - \sigma_3} .$$

The DT parameter risks the likelihood of tensile reactivation of a fault or fracture on a linear scale from 0 ($\sigma_1 = \sigma_n$) to 1 ($\sigma_3 = \sigma_n$). In contrast to the slip tendency parameter (ST), DT incorporates no material or fault properties. Considering the likelihood of tensile failure, DT only relates the fault orientation to the imposed stress field (HILLIS & NELSON, 2005).

2.3.2 Theory of slip tendency

The slip tendency parameter based on Amontons' Laws was introduced by Morris et al. (1996) to identify potentially sliding faults in a given stress field. Morris et al. (1996) defined that the T_s is the equivalent to the coefficient of static friction on a particular fault surface under the assumption that this fault is cohesionless and under normal pore fluid pressure conditions:

Equation 2.23
$$T_s = \mu_{static} = \frac{\tau}{\sigma_n}.$$

Sliding occurs when the T_s value is greater than the assumed critical coefficient of static friction (μ_{crit}) on the fault surface. μ_{crit} is defined as the frictional threshold to be overcome at the instant of sliding:

Equation 2.24
$$T_s > \mu_{crit} = Sliding.$$

In this study a normalised slip tendency value (ST) is used. Herein the ratio of shear stress to normal stress is normalised by the critical coefficient of static friction:

Equation 2.25
$$ST = \frac{\tau}{\mu_{crit}\sigma_n}.$$

A higher ST implies that the state of stress along a given fault surface is closer to failure and that consequently the likelihood of slip is higher. Slip occurs when ST becomes larger or equal to 1. ST as defined in equation 2.25 does not include variations in frictional properties of fault surfaces and is calibrated against the coefficient of static friction of cohesionless faults. The concept of ST is a conservative approach since it overestimates the slip tendency of faults, which are cemented or have an apparent cohesive strength (Streit & Hillis, 2004).

In a more complex scenario, failure depends additionally on the faults cohesion and the pore fluid pressure:

Equation 2.26
$$\tau = C + \mu_{static}(\sigma_n - P_f)$$

Equation 2.27
$$T_s = \mu_{static} = \frac{(\tau - C)}{(\sigma_n - P_f)}$$

Equation 2.28

$$ST = \frac{(\tau - C)}{\mu_{crit}(\sigma_n - P_f)}$$

P_f magnitude of pore fluid pressure

Several studies have demonstrated that cohesion along fault surfaces decreases as subsequent slip occurs and that neglecting cohesion along pre-existing fault surfaces is a reasonable simplification (BRACE & KOHLSTEDT, 1980; KRANTZ, 1991; TWISS & MOORES, 1992; ZOBACK & HEALY, 1992). This reduces equation 2.28 to:

Equation 2.29

$$ST = \frac{\tau}{\mu_{crit}(\sigma_n - P_f)} .$$

In this study the concept of ΔST is also used. This allows determination of the qualitative likelihood of dynamic reactivation of fault surfaces under changing stress conditions. ΔST describes whether a fault surface reaches a state more or less favourable to failure due to a tectonic event:

Equation 2.30 $\Delta ST = ST_2 - ST_1 .$

ST_2 slip tendency after tectonic event
ST_1 slip tendency before tectonic event

2.3.3 Theory of Coulomb failure stress

The dynamic approach of Coulomb failure stress change (ΔCFS) was introduced by REASENBERG & SIMPSON (1992) to investigate possible earthquake interactions. This approach has become a commonly used tool to analyse coherence in changes in stress conditions and earthquake occurrence (e.g. KING et al., 1994; STEIN et al., 1997). It can be used to estimate whether a fault surface is transferred closer to shear failure due to an earlier tectonic event in its vicinity, since the first event changes the state of stress in the surrounding material. The ΔCFS concept allows investigation of whether this stress change can trigger or delay tectonic events on other fault surfaces. It is applicable for both initial shear failure on an incipient fault plane and for fault reactivation and can be calculated for optimally oriented planes as well as pre-existing fault surfaces.

The static Coulomb failure stress (CFS; σ_c or CFF) based on the Coulomb failure criterion, with pore fluid pressure being considered, was defined by REASENBERG & SIMPSON (1992) as:

Equation 2.31
$$\text{CFS} = \tau - \mu(\sigma_n - P_f) - C.$$

Under the assumption that the cohesion and the coefficient of friction remain constant, the change in Coulomb failure stress on a fault surface caused by stress perturbations induced by a nearby tectonic event can be calculated as:

Equation 2.32
$$\Delta \text{CFS} = \Delta \tau - \mu(\Delta \sigma_n - \Delta P_f).$$

When neglecting a change in pore fluid pressure during the tectonic event, the change in Coulomb failure stress is only dependent on the change in shear and normal stress on the fault surface. This reduces equation 2.32 to:

Equation 2.33
$$\Delta \text{CFS} = \Delta \tau - \mu(\Delta \sigma_n).$$

If the absolute values of the stress tensor are not known, the relative change in Coulomb failure stress is generally calculated for rock volumes where no initial stresses are present. This calculation is based on information about the rupture geometry of the previous earthquake and the slip direction of the subsequent earthquake (HARRIS, 1998). Given an unstressed rock, the determination of the change in shear and normal stress magnitudes, acting on a particular surface unit for a subsequent earthquake, is not trivial. The change of the normal stress magnitude is given as a scalar $\Delta |\sigma_n|$. This however provides no information whether the normal stress acting on a surface unit is compressive or tensile. Therefore, $\Delta |\sigma_n|$ has to be normalised relative to the unit normal vector of each particular surface unit. The change in shear stress magnitude of each surface unit is also given as a scalar $\Delta |\tau|$ and must be examined in slip direction of a subsequent earthquake on this particular surface unit. Therefore the slip direction of the subsequent earthquake has to be predefined for each surface unit. The disadvantage of this method becomes evident for cases with numerous arbitrarily oriented surface units for which the slip direction and the type of normal stress have to be determined prior to the ΔCFS calculation. This method is typically used to investigate crustal strike-slip fault zones, where simple rupture geometries are assumed (e.g. HUBERT-FERRARI et al., 2000; CIANETTI et al., 2004; LIN & STEIN, 2004). In this context ΔCFS is calculated for optimally oriented fault surfaces or fault surfaces of a known earthquake sequence in slip direction of each particular subsequent event.

If the absolute values of the stress tensor within a rock volume are known, the relative change in Coulomb failure stress can be easily calculated. Under the convention that compressive stresses are positive, ΔCFS for arbitrarily oriented surface units in an arbitrary stress field can be calculated as:

Equation 2.34 $$\Delta\,CFS = \left(\tau_{postEQ} - \tau_{preEQ}\right) - \mu\left(\sigma_{npreEQ} - \sigma_{npostEQ}\right).$$

τ_{postEQ} shear stress magnitude after previous event

τ_{preEQ} shear stress magnitude before previous event

σ_{npreEQ} normal stress magnitude before previous event

$\sigma_{npostEQ}$ normal stress magnitude after previous event

The algebraic sign of the change in normal stress for each surface unit for a subsequent tectonic event is given by the initial state of normal stress before the previous tectonic event and the state of normal stress after the previous event. Given an arbitrary state of stress, the shear stress acting on each surface unit is defined by the stress tensor and the surface orientation. Therefore, the algebraic sign of the change in shear stress is also given by the initial state of shear stress before the previous tectonic event and the state of shear stress after the previous event.

Positive ΔCFS on a subsequent rupture plane indicates that the plane was brought closer to shear failure and can possibly be triggered by the previous event. Negative ΔCFS indicates that the subsequent plane is now further away from failure. HARDEBECK et al. (1998) considered earthquake triggering due to previous events and defined a threshold based on the observation that only a few events are triggered if the ΔCFS is less than 10 KPa, i.e. only fault surfaces with a positive stress change greater than 10 KPa may be considered to be triggered by a previous tectonic event.

2.3.4 Modelling of fault reactivation

Adequate prediction of the fault reactivation potential requires detailed information on the rheological properties of the faulted rock and the spatial orientation and magnitudes of the acting stress field. Furthermore, comprehensive knowledge of the mechanical properties of the fault surface and the fault geometry is needed. Therefore, a sophisticated method capable of integrating the broad parameter spectrum is needed to evaluate fault reactivation potential. In this study, the method of 3D finite element modelling is used. 3D finite element modelling is a flexible and cost effective method for solving complex mechanical problems. Using the finite element approach, a mechanical problem can be addressed at various resolutions depending on the available data, the problem scale and available computational infrastructure.

To evaluate the reactivation potential, both static and/or dynamic approaches can be used. The static approach used here simulates the 3D stress state by deforming a continuous volume using appropriate boundary conditions. Discrete fault surfaces are not present (or are inactive) during the simulation. After the numerical analysis, the modelled stresses are mapped onto discretised surfaces representing the faults and parameters such as slip tendency or CFS calculated for each surface segment (facet). The dynamic approach used in this study differs from the static one by implementing the faults as frictional contact surfaces, which may become active during the simulation. Application of the boundary conditions causes deformation slip may occur along the fault surfaces. Any slip will result in local perturbations of the stress field. The dynamic approach assumes that continuous slip (fault creep) is possible along the fault surfaces, whereas the static one does not. Similarly to the static approach, a discretised model of the fault surfaces can be used to calculate contact forces acting on each surface unit.

FINITE ELEMENT MODELLING OF THE INTRA-PLATE STRESS STATE

This chapter focuses on the basic concepts of the finite element method (FEM) and the application using the commercial software package ABAQUS™. This includes discussion of:

- Basic principles of the FEM
- Discretisation of the continuum
- Implementation of fault surfaces
- Rheological models
- Multi scale modelling
- Modelling the intra-plate stress state

3.1 The finite element method (FEM)

The finite element method (FEM) is a matrix algebraic method developed to solve partial differential equations (PDEs) in order to calculate the linear or non linear response of continuous physical systems to applied boundary conditions (i.e. loads). The basic principle of the FEM is the division of this physical system into a number of simply shaped continuous sub-domains (i.e. finite elements) in order to find approximate solutions for the PDEs. This so-called discretisation may enable the solution of a PDE system at discrete points in the model space in cases where an analytical solution is impossible to obtain. In order to provide a unique solution and convergence of the equation system, appropriate boundary conditions have to be defined. The simple shape of each finite element is defined by its corner nodes which may be shared with adjacent elements, depending on their spatial position in the model space. All nodes and elements (i.e. entities) are uniquely defined in the model space by their identifier. The entirety of all entities defines the so-called finite element mesh (fe-mesh) which is the discretised representation of the continuous problem space.

Once the model space is discretised, a set of linear equations of the unknown field variable is approximated for each element. The most common approach for problems in continuum mechanics is the so-called displacement method in which the nodal displacement $[\bar{u}(x)]$ is the unknown field variable. Solving the problem for the unknown field variable requires the equilibrium of forces, internal continuity and a constitutive relationship with respect to material behaviour. Through these constitutive laws, stresses are related to strain and hence nodal displacements. The equation of equilibrium can be derived from the equation of motion for small movements which can be written (in index notation) as:

Equation 3.1
$$\frac{\partial \sigma_{ij}}{\partial x_j} + \rho\, B_i = \rho\, a_i\,,$$

where ρ is the density, B_i is the body force and a_i is the acceleration. For the tectonic problem investigated in this thesis, accelerations can be assumed to be negligible since instantaneous displacements are low and the equations of static equilibrium are applicable (RANALLI, 1995). The resultant of all body and surface forces as well as the resultant moment about any axis has to vanish in the state of equilibrium. Therefore, the equation of equilibrium of forces acting in the x-direction can be written as:

Equation 3.2
$$\frac{\partial \sigma_{ij}}{\partial x_j} + \rho\, B_i = 0\,.$$

For perfect linear elasticity, the strain tensor ε_{kl} is proportional to the stress tensor σ_{ij}. This relationship can be expressed as:

Equation 3.3
$$\sigma_{ij} = C_{ijkl}\varepsilon_{kl}$$

and is termed Hook's law. C_{ijkl} represents the elasticity tensor containing 81 components. The number of components reduces to 36 since both the stress tensor σ_{ij} and the strain tensor ε_{kl} are symmetric. Substituting equation 3.3 into equation 3.2, the equation of the equilibrium of forces can be written as:

Equation 3.4
$$\frac{\partial}{\partial x_j}\left[C_{ijkl}\cdot\varepsilon_{ij}\right] + \rho\, B_i = \frac{\partial}{\partial x_j}\left[C_{ijkl}\frac{1}{2}\left(\frac{\partial u_i}{\partial x_j} + \frac{\partial u_j}{\partial x_i}\right)\right] + \rho\, B_i = 0\,.$$

After each element has been approximated into a set of linear equations, the entirety of these approximation functions are assembled into a global equation of motion, which enables the solution of the numerical problem. The matrix algebraic formulation of the fundamental equation of motion can be expressed as:

Equation 3.5
$$\bar{F} = \bar{M}\frac{\partial^2 \bar{u}}{\partial t^2} + \bar{C}\frac{\partial \bar{u}}{\partial t} + \bar{K}\bar{u}\,,$$

where $\overline{M}$ is the mass matrix, $\overline{C}$ is the damping matrix and $\overline{K}$ is the stiffness matrix. In this thesis, the finite element analysis (FEA) is restricted to static and quasi static problems, neglecting accelerations and velocities. Therefore equation 3.5 is reduced to the static equilibrium equation:

Equation 3.6 $$\vec{F} = \overline{K}\vec{u}.$$

The equation of static equilibrium can be inverted to solve for the unknown displacements once the stiffness matrix is known. Subsequently, the stress and strain tensors can be derived.

Two basic principles, the Eulerian and the Lagrangian formulation can be followed to solve the constitutive equations within the fe-mesh (e.g. ALTENBACH & ALTENBACH, 1994). In the Eulerian approach, material properties or field variables migrate through the fe-mesh which is not deforming during the analysis. This prevents numerical instabilities due to excessive distortion of the fe-mesh. However, because the continuum is not explicitly defined, boundary conditions and boundary layers are difficult to trace or redefine. In the Lagrangian approach the material properties are explicitly defined for each element and the fe-mesh deforms during the analysis. The advantage of the Langrangian approach is the traceability of boundary conditions and boundary layers of the sub-domains. The disadvantage of the Lagrangian approach is the explicit definition of the continuum, which leads to excessive distortion of the fe-mesh and numerical instability when addressing large deformation problems. The ABAQUS™/Standard implementation of the Lagrangian approach for structural stress/ displacement analyses is the FEM method used in this thesis.

The major advantage of FEM is that the approximation functions do not have to be differentiable. This implies that each individual element can be assigned unique material properties e.g. when field dependant properties are used. Furthermore, complex geometries can be studied since element shape and resolution can be chosen nearly arbitrarily.

The focus of this thesis is the application of the FEM in a geomechanical context. To give a description of the complete mathematical background of the FEM is beyond the scope of this thesis. Hence, only the fundamental concept and the basic equations have been outlined. For a complete mathematical description of the FEM the reader is referred to ZIENKIEWICZ & TAYLOR (1994, Vol. 1 & 2) and ZIENKIEWICZ et al. (2005).

Applications of the FEM in earth sciences are summarised in RAMSEY & LISLE (2000).

The numerical models presented in this thesis are solved using the standard routine of the commercial finite element software package ABAQUS™ [1] by DS Simulia®, version 6.6-1. The discretisation of the model domains was performed using the commercial finite element software package HyperWorks® [2], versions 7 & 8 by Altair Engineering Inc. The modelling results obtained from ABAQUS were analysed and visualised using, the freely available, GeoMoVie by Peter Connolly, based on OpenDX [3]. For the visualisation of modelling results in profile and map view Golden Software inc. Surfer™ [4], version 8 was used.

(1)http://www.simulia.com/
(2)http://www.altair.com/
(3)http://www.opendx.org/
(4)http://www.goldensoftware.com/

3.1.1 Mesh quality

The accuracy of the FEA is directly dependant on the discretisation resolution and quality. The behaviour of the discretised continuum, for example the field variable $\bar{u}(x)$ is described by a set of linear equations $\bar{u}_\alpha = \alpha_0 + \alpha_1 \bar{u}$, in which α indicates the approximation coefficients. Therefore, a fine discretisation is required in model regions where high gradients of $\bar{u}(x)$ occur (Figure 3.1). Fortunately, the FEM permits to discretise the model space using an irregular mesh which leads to the advantage that the resolution can be locally refined in order to minimise the numerical error.

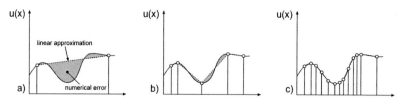

Figure 3.1: Sketch demonstrating the principle of discretisation. The numerical error of the observed field variable (in this case $\bar{u}[x]$) can be minimised by increasing the discretisation resolution stepwise from a) to c).

The accuracy and efficiency of the FEA can also be negatively affected by poor element quality. For example, tetrahedral elements diverging from the equilateral shape can induce so-called stress concentrators or lead to a divergence of the solution.

Figure 3.2 shows a compilation of geometrically distorted tetrahedral elements compared to the desired equilateral shape. In general, the geometric interpretation of the element shape provides a useful rule of thumb to estimate the element quality. Moreover, ABAQUS™ monitors the element quality during the user input and sub-domains can be re-discretised if necessary.

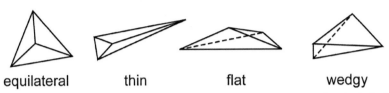

equilateral   thin   flat   wedgy

Figure 3.2: Compilation of geometrical distorted tetrahedral elements compared to the desired pseudo equilateral shape.

3.1.2 Contact problem – modelling of discrete fault surfaces

In this study, the reactivation potential of pre-existing faults under an approximation of the present-day stress field is investigated. In this section, their implementation in the fe-mesh is described. In general, faults can be represented as zones of divergent physical properties or internal boundaries within the discretised continuum which have to be physically defined in the fe-mesh. Thus, faults can be included either as a sub-domain consisting of weaker or stiffer material or as frictional contact surfaces that allow in-surface slip but no separation of the adjacent bodies. In this study both approaches are used, depending on the model scale. The sedimentary infill and the heavily fractured basement of the URG are implemented as sub-domains using weaker (i.e. less stiff) materials. The graben bounding and major intra-graben faults are implemented as frictional contact surfaces. The purpose of a frictional contact surface defined between two sub-domains is to carry relative displacements (i.e. slip) once a pre-defined failure criterion, based on the static coefficient of friction, (μ_s) is exceeded.

To define a contact surface along an internal boundary within the ABAQUS™ model space, the interconnection of the elements of both adjacent sub-domains along this discontinuity has to be abrogated by defining new nodes at one side of the discontinuity. After the duplicate nodes have been created on either side of the discontinuity, those nodes are assigned to each side of the discontinuity and then linked to their corresponding contact elements and areas as either the master or slave surface (Figure 3.3). The master surface is generally chosen to be the stiffer/more coarsely discretised body. During the analysis, the displacements of the slave surface nodes are constrained relative to the master surface nodes by internal boundary conditions derived from the failure criterion defined.

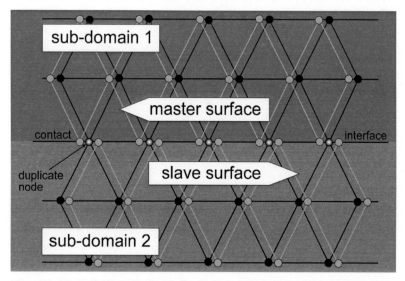

Figure 3.3: 2D sketch illustrating the relative displacement at implementation frictional contact surfaces. Black entities represent the initial spatial position of the fe-mesh, grey entities represent the final spatial position after slip occurred along the contact interface.

In this study, the static coefficient of friction (μ_s) is assumed to be constant and isotropic for all implemented contact surface. A friction value according to Byerlee's law (see section 2.2.1) is assigned ($\mu_s = 0.6$), since no geological evidence exists supporting the assignment of different frictional coefficients to specific faults or fault segments within the URG. Therefore, all fault surfaces have been assigned with the same frictional coefficient. During the analysis, the contact surface status is dependent on the shear and normal stresses acting at each particular point on the contact surface. No relative displacement (i.e. stick) occurs when the actual shear stress at a particular point is less than the critical shear stress to initiate sliding:

Equation 3.7 $\qquad\qquad \tau < \tau_{crit} \text{ and } \tau_{crit} = \mu_s \sigma_n,$

where τ is the actual shear stress, τ_{crit} the critical shear stress and σ_n is the contact pressure (i.e. normal stress). Slip along the contact surface occurs when the actual shear stress reaches the critical magnitude required to initiate sliding:

Equation 3.8 $\qquad\qquad \tau = \tau_{crit} = \mu_s \sigma_n$.

Figure 3.4 demonstrates the generic friction model for the contact surfaces implemented.

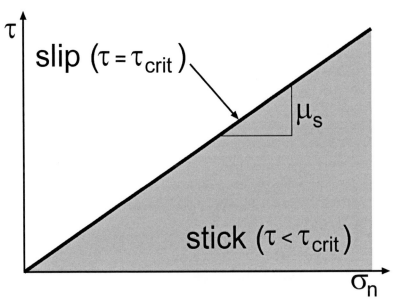

Figure 3.4: Visual representation of the generic friction model used to implement contact surfaces. No relative displacement occurs where the shear stress is insufficient to initiate sliding (grey region). Slip occurs when the actual shear stress becomes critical $\tau=\tau_{crit}$).

3.1.3 Rheological models in ABAQUS™

ABAQUS™ provides a broad variety of rheological models, where the material laws relate the deformation (ε) to the acting stresses (σ) in the body. For example, available rheological models include:

- linear elasticity,
- viscosity,
- plasticity (e.g. Mohr-Coulomb or Drucker-Prager),
- power law creep,
- poro-elasticity or
- combinations of several of these (e.g. visco-elasticity; elastic-plastic).

In general, the rheological behaviour of rock can be approximated by combining three principle rheological elements (e.g. ALTENBACH & ALTENBACH, 1994; RANALLI, 1995): the Hookean elastic element, the Newtonian viscous element and the Saint Venantian plastic element (Figure 3.5). More complex substances can be realised using combinations of these basic elements (e.g. Maxwell substance, Kelvin substance or Bingham substance; e.g. JAEGER et al., 2007).

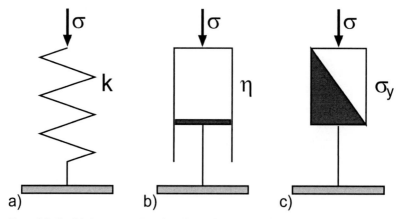

Figure 3.5: Graphical representation of the three principal rheological elements. a) the Hookean elastic element, where k is the constant of proportionality between stress and strain. b) the Newtonian viscous element, where η is the specific viscosity of the material. c) the Saint Venantian plastic element, where σ_y is the yield stress.

Considering the Hookean elastic element (Figure 3.5a), the proportional relation between action (σ) and reaction (ε) is described by Hook's law (equation 3.3), where C describes the elastic tensor. To describe the elastic behaviour of an isotropic material in ABAQUS™, only two independent material constants are necessary to define: the Poisson's Ratio (ν) and the Young's Modulus (E). These constants can be derived from the seismic velocities (v_p and v_s) and the material density (ρ; e.g. BERCKHEMER, 1990) as

Equation 3.9
$$\nu = \frac{1}{2}\left(1 + 1 \Big/ 1 - \frac{v_p^2}{v_s^2} \right) \text{ and}$$

Equation 3.10
$$E = 2\rho\, v_s^2 (1 + \nu).$$

Considering only elastic behaviour, the material response is not time dependant. The results can be scaled over arbitrary time spans.

Considering the Newtonian viscous element (Figure 3.5b), the linear relation between action and first derivation of reaction with respect to time is described by

Equation 3.11
$$\frac{\partial \varepsilon}{\partial t} = \frac{\sigma}{\eta},$$

where η is the specific viscosity of the material. For the viscous material behaviour the response is time dependant. Material properties and deformation rates have to be defined consistently and thus, the results can not be scaled over arbitrary time spans. Visco-elastic material behaviour can be realised by combining a Newtonian element with a Hookean element in serial connection (i.e. Maxwell substance).

Considering the Saint Venantian plastic element (Figure 3.5c), the energy transferred into the solid due to the external action is completely dissipated when the yield stress is reached:

Equation 3.12
$$\sigma < \sigma_y \Rightarrow \varepsilon = 0$$
$$\sigma \geq \sigma_y \Rightarrow \varepsilon = \infty,$$

where σ_y is the considered yield stress. Therefore, the material reaction is irreversible (i.e. perfectly plastic). Elastic-plastic material behaviour can be realised by combining a Saint Venantian element with a Hookean element in serial connection.

3.1.4 Plasticity model (ABAQUS™ *JOINTED MATERIAL)

In order to analyse the localisation of induced plastic failure, a plasticity model is used for the upper crustal regions of the global model. The ABAQUS™ jointed material model provides a simple, continuum model containing a high density of pre-existing parallel discontinuities (i.e. joint surfaces). In addition, the model includes a bulk material failure model, which is based on the Drucker-Prager failure criterion. Due to the higher stability of the numerical solution in combination with the implemented frictional contact surfaces, this bulk material failure model is used for the work presented in this thesis. The bulk material failure criterion is defined by the material cohesion and the friction angle. After the failure criterion is reached, bulk inelastic flow is defined in dependency of the specified dilation angle. The bulk material failure does not consider non-linear plastic flow (e.g. hardening).

3.2 The concept of multi-scale modelling, model evolution and model evaluation

This section describes the concept of multi-scale modelling using ABAQUS™. In addition, the defined model geometries and model boundary conditions as well as the procedure of model calibration and evaluation (i.e. benchmarking) is described in general.

3.2.1 The multi-scale modelling approach

In this study the concept of multi-scale modelling is used in order to simulate the tectonic loading of the URG. ABAQUS™ provides the *submodelling* technique in order to analyse a mechanical problem on different scales and resolutions (ABAQUS™/Standard version 6.6-1 User's Manual, §7.3.1). Appropriate boundaries and boundary conditions for an URG scale model in an intra-plate setting cannot be defined. In contrast, on a regional scale, plate boundaries can be defined as model boundaries since their orientation is known and the displacements or acting forces can more realistically be determined. In general, these boundaries influence large regions (i.e. tectonic plates) and therefore a high resolution on a local scale can not be achieved due to computational limitations. For the area of interest, a high resolution (i.e. fine discretisation) is required to resolve local stress perturbations due to local heterogeneities of the material (i.e. fault zones or change of rock type). By using the concept of multi-scale modelling, an accurate and detailed solution can be obtained on a local scale by mapping a displacement field interpolated from an initial regional scale model onto the arbitrarily chosen boundaries of local scale model. In this study a simplified displacement field is applied onto the boundaries of an URG scale fault block model (see section 4.4.2), which is then used to "drive" two subsequent fault block models of the northern and central URG including the major graben internal and external faults.

3.2.2 Development of model geometries and boundary conditions

In order to simulate the crustal state of stress for the URG region, appropriate model geometries and boundary conditions have to be defined. The models contain the present-day surface topography from a freely available digital elevation model (SRTM30v2 [1]). The Earth's surface acts as a free boundary in all models hence displacements in any direction are not constrained. The Mohorovičić discontinuity (Moho) is also implemented as a discrete surface at which material properties change. The Moho topography implemented is based on a compilation by BARTH (2002; see Figure 1.7). The fault block models of the URG region contain a discrete surface at the boundary between the upper and lower crust (i.e. brittle ductile transition zone; BDTZ) which is extrapolated from the lower boundary of crustal seismicity (LEYDECKER, 2005A) and the Moho topography. Furthermore, the fault block models contain a compilation of first order upper crustal faults (see section 1.3.3), implemented as frictional contact surfaces. All geometries are projected on a sphere with it's radius at sea level defined to be 6,378,135 m to obtain a more realistic stress pattern (see section 3.3.2).

Using the FEM, either displacement and velocity boundary conditions or force and stress boundary conditions can be applied. The displacement method (see section 3.1) solves for the nodal displacements [$\bar{u}(x)$] as the unknown field variable and therefore, nodal constraints at the boundary of the model are more effective to use than force boundary conditions. Using nodal constraints, the displacement at the model boundary is uniquely defined and hence, displacement boundary conditions should be used in order to assure a stable model solution. In this study displacement boundary conditions are defined at the sides and base of the models in the radial, circumferential and meridonal (polar) direction of a spherical coordinate system. In addition to the displacement boundary conditions, gravitational acceleration acting on each element directed towards the Earth's centre is defined.

[1] ftp://e0srp01u.ecs.nasa.gov/srtm/version2/SRTM30/

3.2.3 Model calibration and benchmarking

The FEM uses a discretised approximation of the area or volume of interest. Furthermore, the boundary conditions, the defined model geometries and the distribution of material properties through the volume are largely simplified. Therefore, no unique solution of the problem exists and calibration of the numerical model is necessary in order to achieve a best-fit model. Herein, the process of calibration is the manipulation of the boundary conditions with the aim of matching the simulated and observed dependent variables (ORESKES et al., 1994). In this study, the stress state predicted by the models is calibrated using available in-situ stress estimations and stress indicator data (e.g. World Stress Map, HEIDBACH et al., 2007). For benchmarking of the modelling results, the vertical surface displacements predicted are compared to surface uplift derived from geological and geomorphological data. Furthermore, predicted fault slip rates are compared to available geological and geodetical data. Once the models are calibrated, parameters derived from the calculated stress tensor, such as fracture potential, regime stress ratio (RSR; see section 2.1.2) are analysed in order to describe the possible kinematic behaviour of the URG.

3.3 Modelling the intra-plate state of stress

In order to simulate the present state of intra-plate stress of a specific region different sources of stress have to be considered (see section 2.1.1). Important stress sources are gravitational potential energy differences (ΔGPE) induced by lateral density variations. In addition to obvious stress sources such as lateral boundary conditions (i.e. plate boundary forces) the modelling results are also highly influenced by the model geometry and the assumed initial state of stress. In this section, the rationale for the definition of the initial stress state (i.e. pre-stressing) and the use of spherical model geometries is given.

3.3.1 Defining the initial state of stress – a pre-stressing approach

A fundamental problem in quantitative modelling of geodynamical processes arises because, in nature, any arbitrary volume of rock is in quasi-equilibrium with the in-situ loads and processes such as gravity, tectonic deformation, thermal processes and stress annealing. Consequently, in order to simulate the kinematic behaviour of a tectonic system using a numerical model it is insufficient to consider only the boundary constraints active during the modelled time span. Rather, it is necessary to define an initial state of stress that mimics these natural conditions as closely as possible and then load this "pre-stressed" model with the appropriate kinematic and/or thermal loads. From this basic principle two general problems of the initial model set-up arise.

1. Modelling the kinematics of a geodynamical system requires loading of the model with body forces due to gravitational acceleration. Simply applying a gravitational force to the initial model volume (in which realistically compressible rheologies have been defined) results in a large amount of elastic compaction. However, this behaviour is unrealistic since the real rock formed under continuous gravitational loading and is compacted during deposition and burial.

2. There is a significant underestimation of horizontal stresses when lateral boundary loads are applied over only the modelled time span. This arises because the residual or "un-dissipated" stresses present in the rock due to the burial/deformation history are not honoured. If gravitational loading is applied to the model, an estimation of the residual stresses becomes necessary.

Thus, neglecting either of these problems generally leads to an unrealistic kinematic behaviour of the modelled system. Considering the modelled rock volume as linear elastic material, in the vertical direction, the stress loading (S_V) of the model due to gravitational forces relates to depth, material density and the gravitational constant. This is calculated in absolute magnitudes during the static numerical analysis. In the horizontal direction, the stress loading (S_{Hmean}) of the model is linked to the induced vertical stresses through the Poisson's Ratio (v) by

Equation 3.13 $$S_{Hmean} = \left(\frac{v}{1-v}\right) S_V .$$

S_V vertical stress component magnitude

S_{Hmean} mean horizontal stress magnitude

The coefficient of lateral stress (k) of the calculated mean horizontal stress magnitude and the vertical stress magnitude is only dependent on the Poisson's Ratio and is defined as

Equation 3.14 $$k = \frac{S_{Hmean}}{S_V} = \left(\frac{v}{1-v}\right).$$

Considering a Poisson's Ratio of 0.25, a reasonable number for many crustal rocks (TURCOTTE & SCHUBERT, 2002), the horizontal stress magnitudes are in the order of 1/3 (k = 1/3) of the vertical stress magnitudes. In contrast, stress observations from deep boreholes in general indicate higher k-ratios (e.g. k=1.0, KTB borehole; BRUDY et al., 1997). Thus, by only considering the Poisson's Effect as a source for horizontal stresses within the model volume results in a significant underestimation of horizontal stress magnitudes.

By considering the modelled volume of rock as a porous elastic medium with fluid filled pores, the effective stress principle can be applied. TERZAGHI (1923) stated that the behaviour of a soil or under this consideration a fluid saturated volume of rock is controlled by the effective stress ($\bar{\sigma}$), which is the difference between externally applied stresses (σ) and the internal pore fluid pressure (P_f; e.g. ZOBACK, 2007). In a fluid saturated porous medium the total stress and the pore fluid pressure are acting across any arbitrary plane within the rock volume, whereas the pore fluid pressure only influences the normal components of the stress tensor. Following the Terzaghi definition of effective stress, the total stress relates to depth, density of the water saturated column of rock and the gravitational constant. Therefore, the effective stress can be defined as

Equation 3.15 $\qquad \bar{\sigma} = \sigma - P_f$.

By distinguishing between the vertical and horizontal stress components, the effective vertical stress ($\bar{S}_V$) and the effective horizontal stress ($\bar{S}_{Hmean}$) are defined by

Equation 3.16 $\qquad \begin{aligned} \bar{S}_v &= S_v - P_f \text{ and} \\ \bar{S}_{Hmean} &= S_{Hmean} - P_f. \end{aligned}$

Considering the effective stress principle, the coefficient of lateral stress (k; equation 3.14) has to be redefined as the coefficient of effective lateral stress ($\bar{k}$):

Equation 3.17 $\qquad \bar{k} = \dfrac{\bar{S}_{Hmean}}{\bar{S}_V} = \left(\dfrac{v}{1-v} \right)$.

By rearranging equation 3.17 the effective horizontal stress induced by the weight of the overburden can be calculated as

Equation 3.18
$$\overline{S}_{Hmean} = \left(\frac{v}{1-v}\right)\overline{S}_V .$$

To obtain the total horizontal stress induced by the weight of the overburden in a porous elastic medium, pore fluid pressure has to be added to equation 3.18, which then can be rewritten as

Equation 3.19
$$S_{Hmean} = \left(\frac{v}{1-v}\right)\overline{S}_V + P_f .$$

Equation 3.19 illustrates that considering the effect of the pore fluid pressure increases the horizontal stress magnitudes within the crust significantly. When the effect of the pore fluid pressure is combined with all other contributions to the stress tensor, for example those due to lateral and thermal loads, a stress state valid for the observed kinematic behaviour of the modelled system can be obtained. Unfortunately, using ABAQUS™ coupled pore pressure stress analyses are computational expensive and can not be applied on the large model scales of this study. In this study, pre-stressing is conducted in order to account for the elastic compaction and to approximate the in-situ stress state, assuming hydrostatic pore pressure and residual stresses are present. This pre-stressing is conducted in two steps (gravitational and tectonic pre-stressing). During the gravitational pre-stressing step an initial model stress state is obtained by applying gravity and lateral boundary conditions, where the lateral and basal model boundaries are constraint normal to their respective boundary. During the tectonic pre-stressing step, the prediction of the present-day stress state is obtained by applying gravity and lateral and basal boundary conditions inferred from geological observations. The pre-stressing procedure yields an initial model, which can then be subjected to lateral boundary loads to simulate the time span being modelled.

3.3.2 The near surface horizontal stress paradox

Compilations of in-situ stress estimations show that the horizontal stress components increase relative to the vertical stress component towards the Earth's surface. In addition, near the surface of the Earth, the horizontal stress in general exceeds the vertical stress significantly (Figure 3.6). Furthermore, geological features such as pop-ups, stress-relief buckles and late-formed faults indicate the dissipation of increased near-surface horizontal stresses (ENGELDER, 1993). In contrast to these observations, recent joints can be induced at the surface with just a few hundred meters of uplift and erosion (i.e. unloading; ENGELDER, 1985; HANCOCK & ENGELDER, 1989). This characteristic distribution of crustal stresses can not be explained by only considering the Poisson's Effect and the influence of the pore fluid pressure (equation 3.19). GOODMAN (1980) presented an elastic model where the relative increase of the horizontal stress magnitudes towards the Earth's surface and thus the increase of the coefficient of lateral stress (k) is related to the burial history of the sedimentary rock. GOODMAN (1980) stated that unloading due to erosion tends to increase the value of k towards the surface of the Earth. After ENGELDER (1993), the functional relationship between the coefficient of lateral stress (k) and the change in burial depth can be expressed as

Equation 3.20
$$k_{(z)} = k_0 + \frac{\left[k_0 - \left(\frac{v}{1-v} \right) \right] \Delta z}{z} ,$$

where k_0 the coefficient of lateral stress in the deformation depth z_0 and is not equal to $v/1-v$.

In this study a different approach is used to simulate the relatively increased horizontal stress magnitudes towards the Earth's surface. This observation is also consistent with models approximating the Earth's Lithosphere as a self gravitating spherical shell situated on a massive and unyielding interior (e.g. MCCUTCHEN, 1982; SHEOREY, 1994). SHEOREY (1994) presented a layered, thermo-elastic and spherical model, which relates the observed k-ratios to the Earth's curvature and the baseline at which compaction ceases (i.e. core-mantle boundary). SHEOREY (1994) proposed that the k-ratio within the Earth can be estimated using the following expression:

Equation 3.21
$$k = \frac{v}{1-v} + \frac{\beta EG}{(1-v)\gamma} \left(1 + \frac{1000}{z} \right) ,$$

where β [°C^{-1}] is the coefficient of linear thermal expansion, G [°C/m] is the geothermal gradient and γ is the rock pressure. By only considering the upper crust, equation 3.21 reduces to

Equation 3.22
$$k = 0.25 + 7E_{|GPa|} \left(0.001 + \frac{1}{z_{|m|}} \right) .$$

In Sheorey's model the predicted stress ratio (k) for the upper crust is dependant only on the depth (z) and the Young's Modulus (E), under the assumption that the contribution from topography, geological structure and tectonic forces is negligible.

Figure 3.6 compares coefficients of lateral stress obtained from in-situ stress measurements with the prediction of the Sheorey model for different Young's moduli.

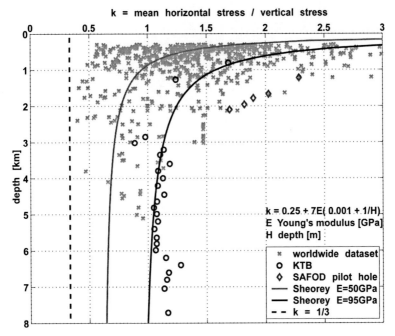

Figure 3.6: Compilations of worldwide in-situ stress estimations (for depths > 300m) including the *Kontinentale Tiefbohrung* in Germany (KTB; BRUDY et al., 1997), the *San Andreas Fault Observatory at Depth* pilot hole (SAFOD; HICKMAN & ZOBACK, 2004). The two solid graphs represent k-ratio predictions of the Sheorey model for different Young's moduli. The dashed line represents the k-ratio prediction of a linear elastic finite element model defined in Cartesian coordinates, assuming a Poisson's ratio of ν=0.25. Note the tectonically increased k-ratios for the SAFOD pilot hole. Figure provided by HEIDBACH et al., submitted.

In order to investigate the effect of a spherical model geometry on the in-situ stress state a generic spherical fe-model at the spatial position of the URG area was constructed (Figure 3.7).

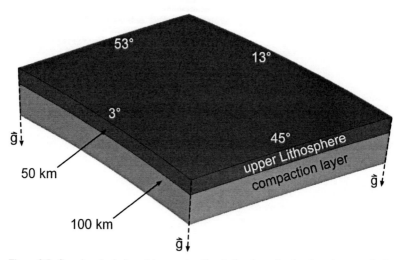

Figure 3.7: Generic spherical model geometry. Gravitational acceleration is acting towards the centre of the Earth. The model dimensions are E3° to E13° (~900 km) and N45° to N53° (~750 km). The total thickness of the model is 150 km.

The model contains two layers with the first layer (50 km thickness) representing the upper Lithosphere and the second layer (100 km thickness) representing an artificial compaction layer that allows the model to compact sufficiently to yield realistic k-values (Figure 3.7). The following equations have to be applied to construct the spherical geometry of a part of the Earth's Lithosphere:

Equation 3.23

$$x' = (r_{(earth)[m]}+z_{[m]})*\cos(x_{(lon)[rad]})*\cos(y_{(lat)[rad]}),$$
$$y' = (r_{(earth)[m]}+z_{[m]})*\sin(x_{(lon)[rad]})*\cos(y_{(lat)[rad]}),$$
$$z' = (r_{(earth)[m]}+z_{[m]})*\sin(y_{(lat)[rad]}).$$

To construct the model geometries, the earth's equatorial radius ($r_{(earth)}$), or semi-major axis is used. The equatorial radius is the distance from the Earth's centre to the equator and is defined to be 6378.135 km.

The input data in geographic projection [longitude (x), latitude (y), height$_{[m]}$ (z)] has to be re-projected to that effect, that the spherical geometry of the Earth is defined in a Cartesian coordinate system [x', y', z'] with its centre [0, 0, 0] defined at the centre of the Earth. In order to simulate the gravity dependent state of stress, the model is accelerated gravitationally towards the Earth's centre. The base of the compaction layer is constrained in the radial direction of a spherical coordinate system which is defined at the centre of the Earth. The sides of the model are constrained in the circumferential and meridonal direction of this coordinate system. For this model, typical rheological parameters for rocks within the Earth's crust are used (E=50 GPa, v=0.25, ρ=2700 kg/m^3; TURCOTTE & SCHUBERT, 2002). In this generic elastic model, the elasticity of the artificial compaction layer simulates the effect of the Earth's interior towards the baseline at which compaction ceases (i.e. core-mantle boundary). Therefore, a sensitivity analysis was conducted for the value of the Young's modulus of the compaction layer. Excessive elastic compaction leads to a significant overestimation in horizontal stress magnitudes. When the Earth's interior is assumed to be rigid, the effect of the spherical geometry is negligible (Figure 3.8; modified after HEIDBACH, et al., submitted).

For a 100 km thick compaction layer with an assumed Poisson's ratio of v=0.25 an elasticity of 750 GPa yielded the best accordance to the predictions of the Sheorey model. It is important to note, that in a thermo mechanical model reasonable values for the elasticity of the Earth's interior could be considered because processes like thermal expansion would counteract excessive elastic compaction (e.g. SHEOREY, 1994).

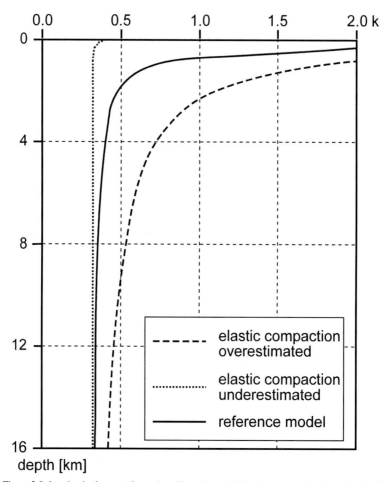

Figure 3.8: k-ratios in the crust for various Young's moduli for the compaction layer during the sensitivity analysis. A relatively low value for the Young's modulus (100 GPa) leads to an overestimation of the horizontal stress magnitudes in the crust. A relatively high value for the Young's modulus (1000 GPa) leads to no significant increase of the horizontal stress magnitudes towards the Earth's surface.

By combining the spherical model geometry with the concept of pre-stressing (see section 3.3.1), a realistic in-situ state of stress can be approximated for the upper crust using elastic rheologies within a finite element model. Figure 3.9 compares the in-situ stress predictions of different modelling approaches using an identical fe-mesh and identical rheological properties.

In contrast to the Sheorey model (equation 3.22), this generic model shows no dependency of the horizontal stress magnitudes predicted at greater depths to the Young's modulus within the crust. The k-values predicted by the FEA are most sensitive to the initial state of stress assumed

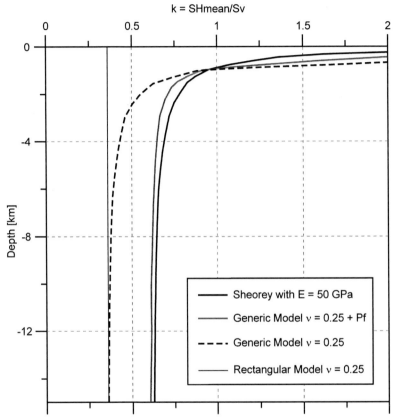

Figure 3.9: Various in-situ stress predictions for the upper crust using the generic model geometry compared to the predictions of the Sheorey model. The elastic parameters of all models are identical (E=50 GPa, v=0.25). Additional pre-stressing of the model yields more realistic horizontal stress magnitudes (e.g. initial pore fluid pressure). The results of a re-projected Cartesian model excluding pre-stressing demonstrate that this approach is insufficient to simulate in-situ stress magnitudes within the crust.

SIMULATING THE IN-SITU STATE OF STRESS OF THE UPPER RHINE GRABEN (REGIONAL MODEL)[1]

4.1   Introduction

The main focus of this regional modelling study (i.e. global model; see section 3.2.1) is to provide boundary conditions for the two regional scale models of the northern and central URG (i.e. sub-model A and B) and to simulate the present-day stress field and the kinematics of the regional scale neotectonic deformation in the URG region. For this purpose, first order sources or boundary conditions for the present-day stress state had to be defined and evaluated. The present-day stress field in the URG region can be related mainly to plate boundary forces (i.e. African-Eurasian convergence and differential gravitational loading at the Mid-Atlantic ridge; e.g. GÖLKE, 1996), the internal structural composition of the model area (e.g. lateral density and thickness variations and the pre-existing fault pattern) and to the inherited loading history of the system (see section 3.3.1). The contemporary tectonic behaviour of the URG has been previously studied in the 1960s and early 1980s (e.g. SEMMEL, 1968; ILLIES & GREINER, 1976; MONNINGER, 1985). More recently, European network projects such as PALEOSIS (PALEOSIS, 2000), ENTEC (CLOETINGH & CORNU, 2005; CLOETINGH et al., 2006) and EUCOR-URGENT (EUCOR-URGENT, 2007) have focused on the interaction between recent surface processes and crustal deformation within the ECRIS. However, the present-day kinematics within the in-situ stress field of the URG is only understood for local areas and only a limited data-set of stress indicators and in-situ stress measurements is available. Furthermore, the factors influencing the kinematic behaviour and the evolution of the URG remain under discussion (e.g. DÈZES et al., 2005; MICHON & MERLE, 2005). Given these limitations, the method of finite element analysis can provide constraints on the kinematic behaviour and estimates of in-situ stress magnitudes and orientations of a geological system (e.g. of the West Netherlands Basin by DIRKZWAGER, 2002). The method of FE modelling is suitable for analysing the tectonic behaviour at various scales ranging from plate-wide models (e.g. GÖLKE & COBLENTZ, 1996, JAROSIŃSKI et al., 2006) to local scale models. Previously the method was used to analyse the tectonic behaviour in sub-regions of the southern URG addressing in particular faulting behaviour (LOPES CARDOZO, 2004; CORNU & BERTRAND, 2005A; 2005B). The structural and depositional evolution of the entire URG

---

[1] This chapter is partly based on the publication: Buchmann, T.J. & Connolly, P.T., 2007. Contemporary kinematics of the Upper Rhine Graben: A 3D finite element approach. Global Planetary Change 58:287-309.

during the Tertiary was investigated using 3D thermo-mechanical modelling (Schwarz & Henk, 2005; Schwarz, 2006). In contrast, this study uses the method of 3D FE modelling to analyse the recent kinematics of the URG within the present-day stress field.

The URG as an abandoned Cenozoic continental rift is superimposed on the structural complex and fragmented Central European Crust. This fragmentation is the result from Variscan and Late-Variscan tectonics (Figure 1.2 and Figure 1.3). Furthermore, the development of the URG was accompanied by magmatic activity and uplift of the crust/ mantle boundary (e.g. Illies, 1974b; Ziegler, 1990, Ziegler,1994; Ziegler & Dèzes, 2006; Figure 1.7). This mantle uplift, probably related to passive buckling of the lithosphere and elevated heat flow in response of Alpine collisional forces, is thought to be ongoing since the Middle Eocene and causing contemporary upper crustal deformation (e.g. Illies, 1981; Monninger, 1985). In this modelling study the contribution of the inherited crustal structure and the far-field stress loading to the present stress state of the system is evaluated. Therefore, the elements and geometric structures required to reproduce the tectonic behaviour and to quantify the contemporary crustal stress field of the present-day URG are implemented. The 3D FE models described herein were calibrated with available in-situ stress data. The best-fit model results are used for benchmarking against the regional uplift pattern, available local fault slip data and the distribution of areas indicating increased recent tectonic activity. The best-fit model results were subsequently used to load more complex local scale models (i.e. sub models; see section 3.2.1) investigating the reactivation potential of second and third order fault structures in more detail. In this section the construction, evaluation of boundary conditions and the loading procedure and calibration of the global model is described. The modelling results are also presented and discussed. This chapter is partly based on Buchmann & Connolly (2007).

4.2   Construction of the global model

The global 3D Finite Element model (FEm) has the entire URG located in the centre surrounded by the adjacent tectonic units. The model contains three lithospheric layers representing the upper mantle, the lower crust and the upper crust consisting of shoulder and graben units (Figure 4.1).

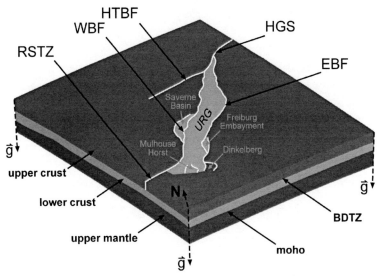

Figure 4.1: Oblique view of the spherical geometry of the URG global model. Gravitational acceleration is acting towards the centre of the Earth. The model dimensions are approximately 450 x 375 km, and the total thickness of the model is approximately 60 km. The model comprises three lithospheric layers and a simplified fault model (white lines) of first order fault surfaces implemented as frictional contact surfaces (see section 3.1.2). HTBF, Hunsrück Taunus Border Fault; WBF, Western Border Fault; RSTZ, Rhine Saône Transfer Zone; HGS, Hessian Graben System; EBF, Eastern Border Fault.

A varying thickness of the upper crustal units is implemented in order to analyse the influence of gravitational potential energy differences (ΔGPE) on the present state of stress in the URG area. Furthermore, the lower crust and the upper mantle were implemented to apply an unevenly distributed load on the base of the upper crust to simulate the effect of an uplift of the crust/mantle boundary on the state of stress affecting the URG area. The geometric basis of this 3D URG FEm was an Earth model including first order fault surfaces, the surface topography, the topography of the brittle ductile transition zone (BDTZ) and the Moho topography built in the goCad® software package. This Earth model was then discretised using Altair® HyperWorks® and re-projected to a spherical coordinate system in order to honour the observed relative increase of horizontal stresses towards the Earth's surface (see section 3.3.2). The surface topography was implemented using the SRTM30 data-set and a surface representing the crust/mantle boundary (i.e. Moho) was implemented based on the integration of seismic reflection and refraction data (BARTH, 2002; Figure 1.7). The

BDTZ is extrapolated from the lower boundary of crustal seismicity (LEYDECKER, 2005A) and the Moho topography.

Lateral depth dependent density variations were assigned to the lower crust and the shoulder and graben units in order to account for the complex Variscan structure of the model area (Figure 1.2 and 1.3).

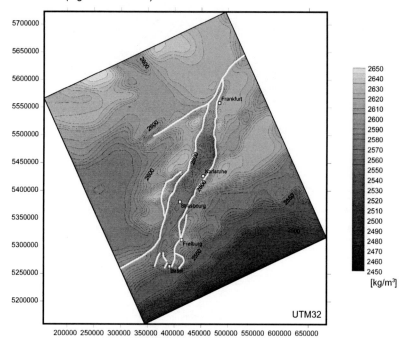

Figure 4.2: Horizontal section through the model at a depth of 2 km illustrating the lateral density variations implemented. In general a decrease in assigned densities from N to S is evident. Relatively high densities were assigned for the Variscan units in the northern and central part of the model (light colours), whereas relatively low densities were assigned to the sediments of the Molasse Basin in the South of the model and to the sedimentary fill of the URG (dark colours).

These density variations were derived from Bouguer Anomaly data (ROTSTEIN et al., 2006; PLAUMANN, 1987) and were assigned as field dependent variables to the nodes of the model (Figure 4.2). Based on this nodal field the unique density at each element integration point was interpolated linearly during the analysis.

For the URG global model first order tetrahedral elements were used. Specifications on the element types of the various units are given in Table 4.1. The lower crust and the lithospheric upper mantle were implemented in order to apply an unevenly distributed vertical load as a function of the Moho topography on the base of the upper crust. This vertical load is induced by gravitational unloading of the upper mantle (see section 4.3).

Model Unit	Element Type	~Element Size [m]	~Quantity
Graben	C3D4	1500	266,000
Shoulders	C3D4	1500-4500	843,500
Lower crust	C3D4	1500-8000	246,000
Upper Mantle	C3D4H	8000-15,000	25,500

Table 4.1: First order tetrahedral elements were used for the discretisation of the various model units. For the upper mantle hybrid elements with uniform pressure were used. A maximum resolution of 1.5 km was achieved within the graben and along the frictional contact surfaces implemented. Average element size and quantity are presented for each model unit.

The compilation of first order upper crustal faults (i.e. border faults) penetrating the entire upper crust was implemented as frictional contact surfaces using Coulomb friction and a typical coefficient of static friction (μ_s) of 0.6 (e.g. TURCOTTE & SCHUBERT, 2002; Figure 4.1). A parametric study of the effect of the friction coefficient assigned was conducted, in which the coefficient of static friction was varied between 0.2 and 0.8. This showed that the influence of the frictional coefficient on fault slip magnitudes and stress orientations was predicted to be small, since the implementation of a calibrated stress state leads to higher normal stresses acting on the individual fault surfaces and a less critically stressed crust. Furthermore, the model area is characterised by a significant horizontal stress unisotropy (see Figure 4.9 & .10) and local re-orientations of the stress field are less likely to occur. No geological evidence exists supporting the assignment of different frictional coefficients to specific faults or fault segments within the URG. Therefore, all fault surfaces were assigned the same frictional coefficient. The listric geometries of the border faults are inferred from their surface traces and from dips estimated from 2D seismic sections (e.g. TIETZE et al., 1979; BRUN et al., 1991; BRUN et al., 1992). The boundary faults generally dip 60° towards the graben centre (PFLUG, 1982). The geometry of the HTBF is more complex than the other faults considered since this structure is part of an inherited Variscan terrane boundary and has been reactivated during both the development of the late-Variscan Saar Nahe Basin (SNB) and during the evolution of the URG (ANDERLE, 1968). The HTBF is a folded fault structure, which is dipping northwards at shallow depths and southwards at greater depth (e.g. MURAWSKI, 1975; ANDERLE, 1968; ANDERLE, 1987).

4.3 Modelling parameters and boundary conditions

For a FEA appropriate boundary conditions have to be defined. In this study nodal constraints were used as boundary conditions at the sides and base of each model for numerical reasons (see section 3.2.1). For a geodynamical analysis the contribution of the far-field loading of a geological system has to be evaluated. The variability of the orientations of the maximum horizontal stress component can be used as an indicator for either the dominance of local or far-field effects on the state of stress observed in a region. In the URG region a relatively homogeneous stress pattern is observed. In general, the maximum horizontal stress component is oriented NW/SE to NNW/SSE and only minor deviations due to local effects are recognised (see Figure 1.12, section 1.4.2). Based on this observation, a simplified displacement field can be used in this study as lateral boundary condition for the URG regional scale model. The model boundaries are oriented perpendicular to the directions of the applied displacement field in order to simplify lateral loading conditions (Figure 4.3). In addition, the magnitude of the lateral loading conditions used to model the present-day state of stress of the URG was inferred from several plate motion models based on GPS data and other modelling studies (DeMETS et al., 1990; KREEMER & HOLT, 2001; NOCQUET & CALAIS, 2004; TESAURO et al., 2005). Due to large uncertainties in the current GPS data set, iterations of the best-fit boundary conditions were necessary. The stresses induced by different lateral boundary conditions were compared to in-situ stress data and the best-fit loading conditions were determined. In order to simulate the recent kinematic behaviour of the URG a time span that was assumed to be sufficiently long to represent the contemporary loading of the system. It was found, based on a sensitivity analysis, that over a time span of 10 ka, lateral boundary conditions of 0.56 mm/a shortening in NNW/SSE-direction and 0.4 mm/a extension in ENE/WSW-direction provide the best-fit to the calibration data (Figure 4.3).

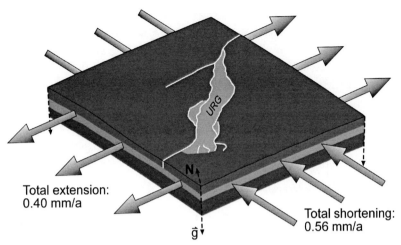

Figure 4.3: Oblique view of the URG global model loading procedure. Arrows indicating the simplified lateral loading conditions applied perpendicular to the model boundaries. For the URG modelling study presented herein, gravitational acceleration and nodal displacements defined at the model sides were used. The upper surface of the models remains unconstrained and acts as a free surface, whereas the base of the model is constrained in the radial direction.

Deformation within the brittle upper crust is described by elastic or elasto-plastic material behaviour. In order to honour the different material properties of the highly fractured graben interior and the less fractured shoulder regions two different rheological units were defined for the upper crust. In general, these two rheological units are separated by frictional contact surfaces representing the graben bounding faults. At the scale of this URG regional model the 2 to 3 km thick sedimentary infill of the graben is neglected. The lower crust and the lithospheric upper mantle are also implemented using elastic material properties. Viscous or visco-elastic material properties were not used for the following reasons:

- The use of viscous materials within a model including topographies requires all model units to be in gravitational equilibrium at the initial step of the model to avoid unrealistic or unobserved surface displacements.
- The non-uniform thickness defined for the upper crust would therefore require assignment of a unique density for each column of rock within the model.
- Furthermore, for the region of the Vosges Black Forest Dome no static gravitational equilibrium can be achieved since elevated surface topography in this region is associated with a distinct culmination of the Moho (Figure 1.7).

Thus, no stable boundary conditions could be found that keep the model in static gravitational equilibrium using viscous or visco-elastic material properties for the lower crust and upper mantle respectively. Therefore, the use of elastic material behaviour for the lower crust and the upper mantle is a necessary simplification that nevertheless leads to a realistic approximation of the upper crustal state of stress. The rheological properties of the units implemented are given in Table 4.2.

Model Unit	Rock Type	Density [kgm^{-3}]	Young's Modulus [GPa]	Poisson's Ratio	Reference
Graben	Sediments, fractured basement	2450-2750	45	0.275	Turcotte & Schubert, 2002
Shoulder	Sandstone, Granite and Gneiss	2550-2750	55	0.275	Turcotte & Schubert, 2002
Lower crust	Anorthosite	2750-2850	83	0.4	Turcotte & Schubert, 2002
Lithospheric Upper mantle	Dunite	3350	150	0.5	Turcotte & Schubert, 2002

Table 4.2: Elastic rheological properties assigned to the 3D FEm units.

For the description of elasto-plastic material behaviour within the upper crust a simplified Drucker-Prager failure criterion is used (see section 3.1.3). A uniform failure criterion is assigned to both the shoulder and graben units. The parameters defining the occurrence of plastic failure are given in Table 4.3.

Model Unit	Friction Angle	Dilation Angle	Cohesion [MPa]
Shoulder & Graben	15°	15°	7.6

Table 4.3: Rheological parameters assigned to the upper crustal units of the model describing the initiation of plastic failure.

During the iteration of the best-fit lateral boundary loads, the initial stress state and kinematic behaviour of the URG could not be reproduced unless the effect of varying crustal thickness and Moho uplift were taken into account. Several authors argue that the formation of the URG was accompanied by Moho uplift, probably related to buckling of the lithosphere and elevated heat flow due to the Alpine collision (e.g. ILLIES, 1974B; ZIEGLER, 1990 & 1994). This mantle uplift has commenced in the Middle Eocene, culminated during the Miocene and is thought being the cause for contemporary upper crustal deformation (ILLIES, 1981; MONNINGER, 1985). In order to implement a varying upper crustal thickness and to simulate the effect of Moho uplift possibly induced by buckling of the lithosphere, the lower crust and the lithospheric upper mantle had to be included in the model. This allowed the application of an unevenly distributed vertical load, as a function of the Moho topography, to the base of the upper crust. This basal boundary condition applied on the base of the upper crust was controlled by using variable densities for the upper mantle. During pre-stressing, the mantle was assigned a density greater than the real value. Then, during the main loading phase the mantle density was reduced to its real value (Table 4.2). The uplift was induced at the Moho as a result of the uplift of the now lighter mantle; i.e. net buoyancy is applied to the base of the upper crust (see section 4.4.1). Iteration of this basal loading condition has shown that the model results are very sensitive to the artificial upper mantle density variations used. It was found that a density 10% higher than the expected mantle density of 3350 kgm^{-3} yielded suitable differential uplift of the shoulder and graben regions for the northern URG area.

4.4 Loading procedure and calibration

The loading procedure of the URG regional model (i.e. global model) comprises the following steps:

- a gravitational pre-stressing step to establish realistic depth related stress magnitudes within the model and to counteract excess elastic compaction,
- a tectonic pre-stressing step to simulate the effect of the long-term deformation history of the URG on the present-day state of stress,
- and an analysis of the short-term (10 ka) deformation pattern of the URG induced by lateral loading of the system.

4.4.1 Gravitational pre-stressing model

In this first step the model is only loaded by gravitational acceleration towards the Earth's centre and is laterally constraint by a displacement field interpolated from the calibrated generic spherical model described in section 3.3.2. In order to achieve a gravitationally loaded initial model, the URG global model was embedded in this model using the ABAQUS™ *submodelling* technique (see section 3.2.1; Figure 4.4).

The interpolated nodal displacements are transferred from the global model (generic spherical model, see section 3.3.2; Figure 3.7) to the sub-model (URG global model). The lateral sides of the URG global model are constrained in the circumferential (E/W) and the meridonal (N/S) direction. The base of the model is constrained in the radial direction. The upper surface remains unconstrained and acts as a free surface. Figure 4.5 illustrates the total amount of elastic compaction in the radial direction of the URG regional model.

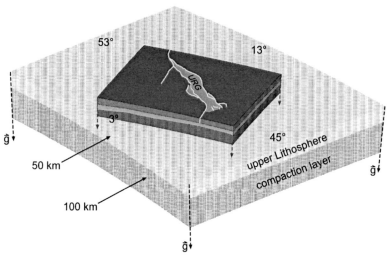

Figure 4.4: Gravitational pre-stressing of the URG global model. The model is loaded only by body forces induced by gravitational acceleration towards the Earth's centre. In order to achieve a model with relatively increasing horizontal stresses towards the Earth's surface, the model is laterally constrained by nodal displacements interpolated from the calibrated generic spherical model of the URG area (see section 3.3.2).

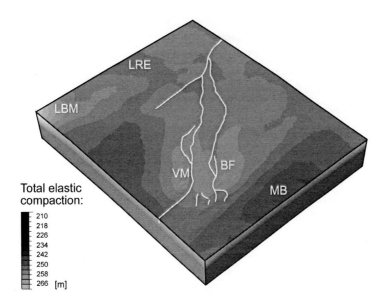

Figure 4.5: Total amount of elastic compaction of the URG regional model after the gravitational pre-stressing step. The distribution of the elastic compaction is a result of the Moho and surface topographies and the lateral density variations implemented in the model (see Figures 1.7 and 4.2). Herein, the amount of elastic compaction is a function of the total mass of the vertical material column. A relatively large amount of elastic compaction is predicted for the region of the London Brabant Massif (LBM) in the north-western corner of the model. This area exhibits the largest crustal thickness in the model and relatively high densities of crustal rocks. Additionally, a relatively increased amount of elastic compaction is predicted for the Vosges Mountains (VM) and Black Forest (BF) situated above the Vosges Black Forest dome where the crustal thickness is lowest. A relatively low amount of elastic compaction is predicted for sedimentary basins such as the Lower Rhine Embayment (LRE) and the Molasse Basin (MB).

The gravity-induced stresses are unique for each point within the URG regional model since variable topographies and lateral density variations are implemented. The gravitational pre-stressing step is conducted in order to establish an initial gravity induced state of stress within the model which accounts for the inhomogeneous composition of the lithospheric layers implemented (e.g. Figure 4.2). Moreover, by defining lateral thickness and density variations using realistic elastic material properties within an un-stressed model leads to unrealistic differential compaction of the simulated rock mass. This induced differential compaction is shown in Figure 4.5.

The second purpose of the initial pre-stressing step is to establish a state of stress within the model, which yields realistic and observable stress magnitudes with depth. Using a linear elastic material behaviour to describe the mechanical response to the applied gravitational loading of the crust results in a significant underestimation of horizontal stress magnitudes since the horizontal loading of the model is linked to the induced vertical stresses only through the Poisson's Ratio (v; see section 3.3.1, Equation 3.1.3). The linear elastic approach neglects other significant contributions to the state of stress observed in a geological system, for example the acting pore fluid pressure and the effect of the compaction history of the modelled rock mass (see section 3.3.1 and 3.3.2). These effects in general lead to a more isotropic state of stress than predicted by a model using linear elastic material properties (see section 3.3.1, Equation 3.19). In order to simulate the contribution of for example the acting pore fluid pressure using linear elastic material properties, the Poisson's Ratio has to be manipulated during the gravitational pre-stressing step since the coefficient of lateral stress (k; see section 3.3.1) is only dependant on this factor. In this study, the necessary increase of the isotropic part of stress tensor was calculated using available in-situ stress determinations in the model area (e.g. RUMMEL et al., 1986). The Poisson's Ratio for the modelled rock volume necessary during gravitational pre-stressing is then given by rearranging Equation 3.14:

Equation 4.1 $$ v = \frac{k}{1+k} . $$

When this "required Poisson's Ratio" is combined with the effects of all other contributions to the stress tensor, for example those due to lateral and thermal loads, a stress state valid for the observed kinematic behaviour of the modelled system can be obtained. It is important to note that after the gravitational pre-stressing step the horizontal stress magnitudes only reflect the effect of differential gravitational loading of the model due to lateral thickness and density variations. Therefore, relatively low horizontal differential stress magnitudes are predicted by this model (Figure 4.6).

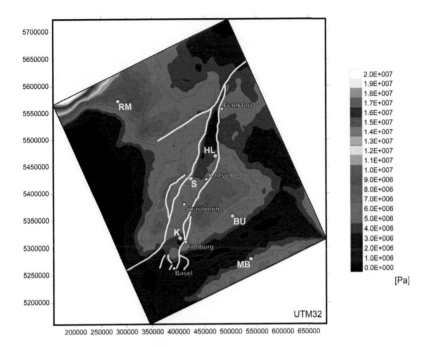

Figure 4.6: Horizontal differential stress magnitudes predicted by the gravitational pre-stress model at a depth of 3000 m. Within the URG area relatively low horizontal stress magnitudes from 1 to 10 MPa are predicted. Relatively high horizontal differential stress magnitudes (up to 20 MPa) are only predicted in the north-western corner of the model where the gradient of the Moho topography is highest. Yellow dots indicate the position of the synthetic wells analysed in the modelled volume for calibration and benchmarking (see section 4.4.3). RM = Rhenish Massif; HL = Heidelberger Loch; S = Soultz sous Forêts; BU = Bad Urach; K = Kaiserstuhl; MB = Molasse Basin.

During the gravitational pre-stressing step a higher density value than the expected density is assigned to the upper mantle. This boundary condition is used to induce an uplift of the Moho along the graben axis as a result of the uplift of the now lighter mantle during the tectonic pre-stressing step; i.e. net buoyancy is applied to the base of the elastic upper crust.

4.4.2 Tectonic pre-stressing model

For the tectonic pre-stressing step the stress tensor obtained for each unique integration point by the gravitational pre-stressing step is used as an initial stress condition. Using this initial stress condition leads to a gravitationally loaded but undeformed model. In the next step this gravitationally loaded model is then deformed during the tectonic pre-stressing step in a lateral sense by implementing a simplified displacement field which is applied to the nodes along the lateral model boundaries. In order to obtain a simplified displacement field for the URG global model the World Stress Map data base (WSM) is used. Based on the direction of the maximum horizontal stress component obtained from in-situ stress determinations in the URG region a simplified orientation field is calculated (Figure 4.7; MÜLLER et al., 2003). The mean orientation value is weighted according to the WSM data quality ranking (A=1, B=0.75 and C=0.5; see:

http://www-wsm.physik.uni-karlsruhe.de/pub/data_details/WSM_quality_ranking.pdf).

Figure 4.8 demonstrates the deviation between the simplified stress field determined and the WSM data points.

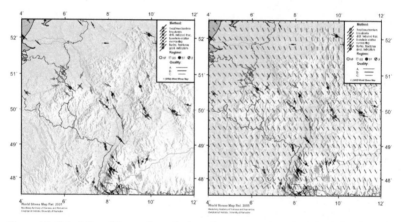

Figure 4.7: Orientation of the maximum horizontal stress component from the WSM data base (left Figure) and the derived smoothed stress field for the modelling area (right Figure; search radius = 250 km, data points 170; provided by O. Heidbach). The smoothed stress field has a mean deviation of +/- 19.2° whereas the mean error of the data points within the WSM data base is +/- 25°.

For the loading of the global model the model boundaries are oriented perpendicular to the inferred simplified stress field. By assuming quasi linear elastic behaviour of the model domain, the inferred orientation of the maximum horizontal stress direction is sub-parallel to the orientation of the maximum horizontal strain direction. Displacement boundary conditions equivalent to 0.56 mm/a shortening in NNW/SSE-direction and 0.4

mm/a extension in ENE/WSW-direction were applied to the lateral model boundaries. During the tectonic pre-stressing step elastic material properties are used. The best fit to the calibration data, which will be discussed in the following section, was achieved after a timeframe equivalent to 200 ka.

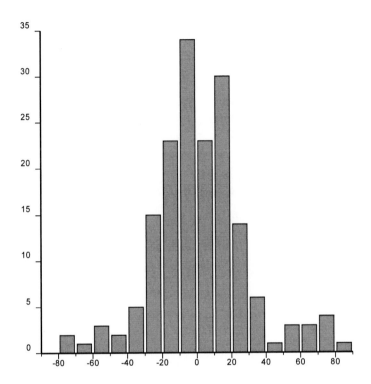

Figure 4.8: Histogram demonstrating the distribution of the deviation between the simplified stress field used to infer the direction of the displacement boundary condition and the WSM data base. The systematic error of SH orientation determinations for C quality data within the WSM data base is 25°. Figure provided by O. Heidbach.

4.4.3 Calibration and benchmarking of the global model using in-situ stress data

The calibration of a numerical model is necessary in order to achieve a best fit model (see section 3.2.3). In the region of the URG global model in-situ stress data is available for two deep hydrothermal well sites. For the calibration of the global model in-situ stress data from the Soultz sous Forêts site (see section 1.4.2, Table 1.1) was used. During the process of calibration the lateral boundary conditions were applied incrementally and the predicted stress magnitudes were compared to the calibration data from Soultz sous Forêts. In order to constrain the process of tectonic pre-stressing not only the present-day state of stress inferred from in-situ stress measurements has to be considered, but it is also necessary to verify that the general tectonic behaviour remained almost constant during the time span considered. For this purpose, palaeo-stress analysis can provide useful information on the duration of the general tectonic conditions affecting a geological system. ANDRÉ et al. (2001) analysed the orientation of quartz veins for different depth intervals from the Soultz sous Forêts site in order to determine palaeo-stress magnitudes or R-ratios respectively (see section 2.1.2). Table 4.4 compares the model prediction with the results of the analysis conducted by ANDRÉ et al. (2001).

Depth Interval	R-ratio (ANDRÉ et al., 2001)	R-ratio model prediction
1400 – 1500 m	0.63	0.68
1625 – 1725 m	0.72	0.75
2075 – 2220 m	0.81	0.81

Table 4.4: Comparison of the stress ratio (R) at the Soultz sous Forêts site derived from a palaeo stress analysis and modelling results after the tectonic pre-stressing step. The palaeo stress analysis indicates that the tectonic regime is characterised by high R values increasing with depth. Note the close fit of the modelling data with the palaeo stress data.

The palaeo stress state inferred by ANDRÉ et al. (2001) shows congruence to the present-day state of stress determined by in-situ stress measurements at the same location both in stress magnitudes and orientations. Therefore, it can be concluded that the Soultz sous Forêts site has been subjected to relatively stable stress conditions since the vein opening occurred. Figure 4.9 compares the modelled stress paths with the different in-situ stress determinations for the Soultz sous Forêts site provided by different authors.

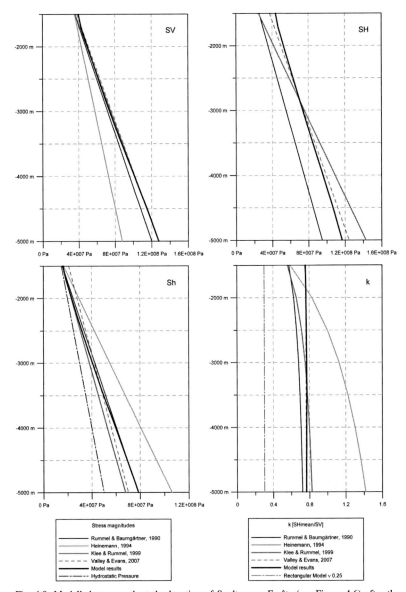

Fig. 4.9: Modelled stress-path at the location of Soultz sous Forêts (see Figure 4.6) after the tectonic pre-stressing step compared to in-situ stress estimations provided by several authors (RUMMEL & BAUMGÄRTNER, 1991; HEINEMANN, 1994; KLEE & RUMMEL, 1999; VALLEY & EVANS, 2007). The calibration of the model is successful within the limits of the applied method, the model resolution and the uncertainties within the calibration data.

At the regional scale, the direction of the maximum horizontal stress component SH in the URG area is assumed to be N155°. This orientation of the maximum horizontal stress direction has been determined using mainly earthquake focal mechanism data from the WSM data base (Figure 4.7). In contrast, in-situ stress determinations from deep geothermal wells in general suggest a local orientation of N170° for the maximum horizontal stress direction at a depth of 3000 m below sea level (e.g. Soultz sous Forêts and Bad Urach; HEINEMANN et al., 1992; TENZER et al., 1992; CORNET et al., 2007).

A general problem exists in exclusively using the in-situ stress data obtained from the Soultz sous Forêts site because of the geological setting of the geothermal reservoir. The reservoir is located in a pervasively fractured granitic body within the central segment of the URG. The N/S orientation of the pre-existing hydraulic fractures within the reservoir body is consistent with the assumed stress fields of both the N/S oriented Eocene shortening and the E/W oriented Oligocene extension phase (ANDRÉ et al., 2001 and references herein). The palaeo stress analysis as well as in-situ stress measurements using hydraulic fracturing tests indicate a local rotation of the maximum horizontal stress direction from a general assumed NNW/SSE to a N/S trend (Figure 4.7). In contrast, maximum horizontal stress orientation determinations using earthquake focal mechanism inversion reveal no local rotation of the stress field at the Soultz sous Forêts site (S_H orientations of N125° +/- 20°; HELM, 1996). The local rotation of the stress field determined might be related to the pre-existing Eocene and Oligocene fabric of the granitic body. However, in-situ stress determinations reveal this structural orientation of the early stages of rifting because of the distinct anisotropy of the volume investigated. A similar structural anisotropy is also found for the Bad Urach geothermal site. Here, the dominant fracture set in the metamorphic reservoir section is also oriented N/S (TENZER et al., 1992) and in-situ stress determinations also reveal a sub-parallel orientation of the maximum horizontal stress direction. It remains uncertain why a systematic rotation of about 15° north of the maximum horizontal stress direction is observed in geothermal wells for a depth interval from 1000 to 5000 m (CORNET et al., 2007). A local scale rotation of the stress field observed for the geothermal reservoir sites cannot be addressed with the model scales used in this study. In contrast, the absolute stress magnitudes inferred at these locations provide the only available constraint about the in-situ state of stress of the system.

Concerning the calibration of the model using in-situ stress data, the following approach is chosen. A numerical model does not provide a unique solution for the problem addressed because of the necessary simplifications used (see section 3.2.3).

For this reason, the quality of the extrapolation of the predicted state of stress provided by the model has to be assured by comparing the modelling results to independent data at a location different to that used as the calibration site. This means that the inferred stress magnitudes from the Soultz sous Forêts geothermal site are used to calibrate the predicted stress magnitudes of the numerical model (Figure 4.9), whereas for the benchmarking of the model in-situ stress estimations from Bad Urach are used (Figure 4.10). The comparison of the predicted stress magnitudes with the stress magnitudes obtained using in-situ stress measurements for the site of Bad Urach indicate that the extrapolation of the state of stress by the model is a valid solution within the limits of the methods and the uncertainties of the reference data.

For further benchmarking of the model, stress data from synthetic wells in the model volume are compared to plate-scale stress estimations and generic models. RUMMEL et al. (1986) investigated hydraulic fracturing data from numerous boreholes and determined an averaged k ratio for the Central European crust, which is nearly constant with depth (H):

Equation 4.2
$$k_H = \frac{S_H}{S_V} \frac{0.27}{H}$$

Equation 4.3
$$k_h = \frac{S_h}{S_V} \frac{0.15}{H}$$

The in-situ stress determinations from the geothermal sites used for calibration and benchmarking as well as the observations of RUMMEL et al. (1986) reveal both k ratios of ~0.8, which implies strike-slip to extensional tectonic regimes for the model area. In contrast, stress estimates from the KTB deep borehole situated in the region of the Oberpfalz to the east of the modelling area indicate a higher k value. In this area, a more compressional state of stress is observed (BRUDY et al., 1997). Figure 4.11 compares the model predictions for various synthetic wells with in-situ stress estimations for the European crust provided by RUMMEL et al. (1986), the KTB deep borehole site (BRUDY et al., 1997) and the prediction of a generic model by SHEOREY (1994; see section 3.3.2). Comparison of the modelled state of stress with the available reference data suggests that the model provides one valid approximation of the state of in-situ stress.

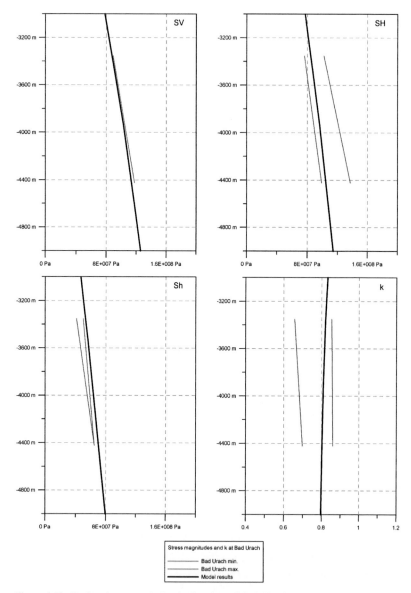

Figure 4.10: Predicted stress-path for the location of Bad Urach compared to in-situ stress estimations provided by TENZER et al., 2001. The modelling results generally show a good fit to the observed data within the limits of the used method and the uncertainties within the benchmarking data used. The location of the Bad Urach site is given in figure 4.6.

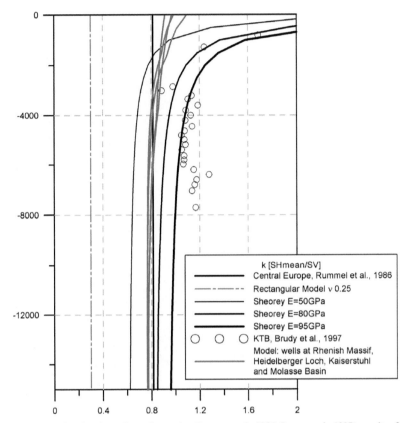

Figure 4.11: k ratios (k_{mean}) from observation (RUMMEL et al., 1986; BRUDY et al., 1997), results of generic models (rectangular fe-model only assuming a Poisson's ratio of 0.25 and SHEOREY, 1994) and predictions by this modelling study. The location of the artificial well sites is given in figure 4.6.

4.4.4   Modelling the short-term deformation pattern of the URG

In order to address the short-term deformation pattern of the URG area, the inferred boundary conditions are applied over a time increment of 10 ka (see section 4.5.4). The vertical surface displacements obtained were compared to uplift data derived from geological and geomorphological data. In addition to the surface uplift predictions, modelled fault slip rates are compared to fault slip rate data available. This comparison is discussed in the following section.

4.5 Modelling results

The purpose of the global model is to obtain an approximation of the present-day state of stress affecting the URG area. The calibrated and benchmarked global model is later used to transfer the applied loading procedure and boundary conditions on subsequent sub-models which aimed at investigating the kinematic behaviour of the URG in more detail (see chapter 5 and 6). In this section, the results from the URG global model study are presented. For comparison, the modelling results are presented for a uniform reference depth of 3000 m below sea level.

4.5.1 Prediction of stress component magnitudes (S_v, S_H and S_h)

In general, the depth distribution of the relative stress component magnitudes is a function of the topographies implemented and the densities assigned. The modelling results suggest that the vertical stress component (S_v) is relatively increased in regions characterised by elevated surface topography (Figure 4.12). Within the URG a general decrease in relative S_v magnitudes from the South to the North is evident.

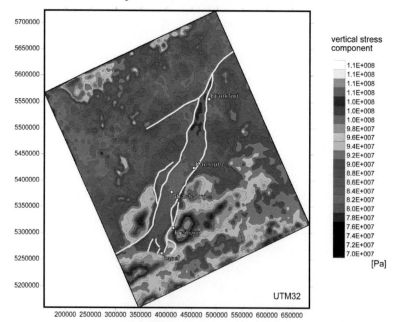

Figure 4.12: Calculated magnitudes of the vertical stress component (S_v) at 3000 m depth. The relative magnitude of the vertical stress is primarily dependant on the surface topography (i.e. overburden). Therefore, S_v is relatively increased for the Rhenish Massif, the Vosges Mountains and Black Forest, the Jura Mountains and the Molasse Basin.

In contrast to the vertical stress component, the pattern of the relative magnitude of the maximum horizontal stress component (S_H) is primarily dependent on the Moho topography. Regions within the Rhenish Massif and Molasse Basin, characterised by an increased crustal thickness, show an increased relative magnitude of S_H (Figure 4.13). Furthermore, the relative magnitude of S_H is dependent on the Young's modulus and the density value assigned. In the vicinity of the frictional contact surfaces the relative magnitude of the maximum horizontal stress component is likely to be decreased due to relative displacement. Within the URG the S_H magnitudes are relatively increased in the central segment, whereas a relative decrease in S_H magnitudes is visible for the northern and southern segments.

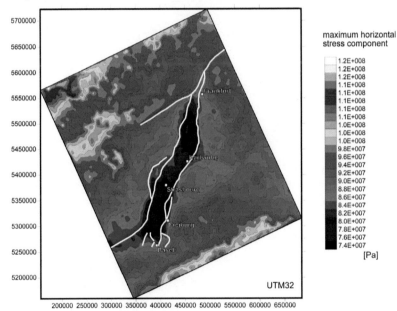

Figure 4.13: Calculated magnitudes of the maximum horizontal stress component (S_H) at 3000 m depth. The magnitude of the maximum horizontal stress is primarily dependent on the Moho topography (i.e. crustal thickness; see figure 1.7). S_H is highest for parts of the Rhenish Massif and the Molasse Basin.

The pattern of the relative minimum horizontal stress component magnitude (S_h) shows a dependency on both elevated surface topography and variations in crustal thickness. Within the URG the S_h magnitudes are also relatively increased in the central segment, whereas a relative decrease in S_h magnitudes is visible for the northern and southern segments. In contrast to the S_H magnitude distribution, S_h is additionally increased for the southernmost part of the URG.

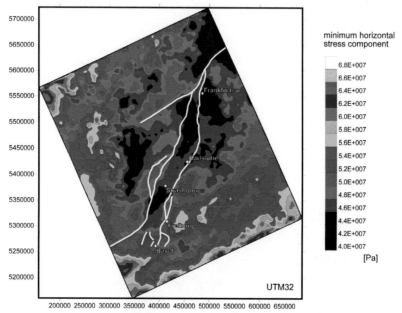

Figure 4.14: Calculated magnitudes of the minimum horizontal stress component (S_h) at 3000 m depth. The magnitude of the minimum horizontal stress shows a dependency on both, elevated surface topography (S_h is relatively increased for the Vosges Mountains and Black Forest and the Jura Mountains) and crustal thickness variations (variation in magnitudes within the Rhenish Massif).

The relative increase of the predicted horizontal stress components S_H and S_h in the NW and SE (Rhenish Massif, Molasse Basin) is a function of the implemented topographies. In general, higher horizontal stress magnitudes are limited to the regions within the model with relatively increased crustal thickness (Figures 1.7. 4.13, 4.14). Hence, increased horizontal stress magnitudes are induced by lateral density variations related to thickness changes of the different lithospheric layers. Lateral density variations in the lithosphere lead to a potential energy difference (ΔGPE) causing a lateral pressure gradient, which is compensated by increased horizontal stress. The effect of the variable thickness of the lithospheric layers on the horizontal stress magnitudes generated is most obvious when comparing Figure 1.7 and Figure 4.13. It is interesting to note that in contrast to the model prediction, kinematic models derived from seismological observations argue that, in general, S_H decreases with distance from the Alps and is lowest in the Lower Rhine Embayment to the northwest of the URG (AHORNER et al., 1983; DELOUIS et al., 1993; PLENEFISCH & BONJER, 1997).

4.5.2 Predicted style of deformation (RSR)

The values of the regime stress ratio (RSR) calculated for the graben and shoulder units show a gradual change in the expected deformation style (Figure 4.15). A transtensional to strike-slip faulting regime is predicted for the Molasse Basin in the southern part of the model (RSR values of 1.0 to 1.3). For the region of the Jura Mountains, the Vosges Mountains, the Black Forest and the Swabian Alp the model predicts a predominantly extensional faulting regime (RSR values of 0.6 to 0.9). For the the Rhenish Massif in the northern part of the model a transtensional to pure strike-slip faulting regime is predicted by the model (RSR values of 0.9 to 1.6; Figure 4.15; see section 2.1.2).

Within the URG the model results show a change in predicted RSR values from an extensional faulting regime in the South to a transtensional faulting regime in the central and northern graben segments. In the vicinity of the Vosges Black Forest Dome the model predicts a higher dominance of the vertical stress component (S_V) and a relative decrease of the maximum horizontal component S_H, which results in lower RSR values. The relative increase in S_V predicted by the model results from the elevated surface topography of this region (Figure 4.12). Coevally, the relative magnitude of S_H is decreased due to the elevated Moho in this region (Figure 1.7 and 4.13).

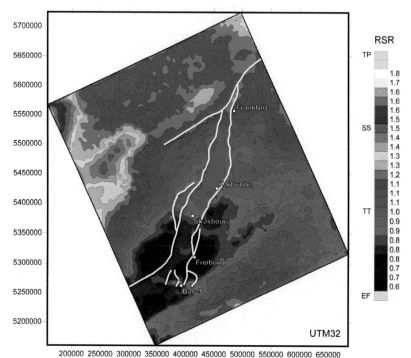

Figure 4.15: Calculated value of the regime stress ratio (RSR; see section 2.1.2) at 3000 m depth. A gradual change in the predicted style of deformation from SE to NW is indicated by a change in RSR value from 1.2 (TT – transtensional faulting regime, Molasse Basin) over 0.6 (EF – extensional faulting regime; Vosges Mountains and Black Forest) to 1.5 (SS - strike-slip faulting regime; Rhenish Massif).

4.5.3 Differential stress and plastic failure prediction

In order to qualify the relative likelihood of brittle failure predicted by the model several parameters need to be analysed. The relative magnitude of the second invariant of the stress tensor (i.e. von Mises stress, σ_M; see section 2.1.2.3; Figure 4.16) is often used to predict possible plastic failure. Relatively increased magnitudes of σ_M are predicted for large parts of the shoulder regions, whereas the relative magnitudes are influenced both by elevated surface topography and crustal thickness. Due to the lower elasticity applied to the graben region, a more isotropic state of stress is predicted for the URG. Within the graben σ_M is relatively decreased in the vicinity of appropriately oriented frictional contact surfaces. A relative increase in predicted σ_M magnitudes is observable in the central segment and for regions in the northern and southern URG segments.

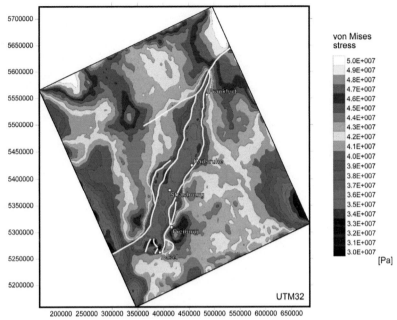

Figure 4.16: Calculated von Mises stress magnitudes (σ_M; see section 2.1.2.3) at 3000 m depth. For the shoulder regions the magnitude of σ_M shows a dependency on both elevated surface topography (e.g. Mainz Basin, Kraichgau Trough, Vosges Mountains and Black Forest) and crustal thickness variations (Rhenish Massif).

The magnitude of the horizontal differential stress (S_{Hdiff}) is an indicator for the anisotropy of the calculated stress tensor in the horizontal direction. Relatively increased magnitudes of S_{Hdiff} are predicted for regions with increased crustal thickness (Rhenish Massif and the Molasse Basin; Figure 4.17). Within the URG S_{Hdiff} magnitudes are relatively increased in the central segment and parts of the northern segment whereas the southern segment is characterised by a more isotropic state of horizontal stress. The relatively increased isotropy in horizontal stress magnitudes predicted by the model is induced gravitationally by the elevated surface topography accompanied by an elevated mantle in the region of the Vosges Black Forest Dome.

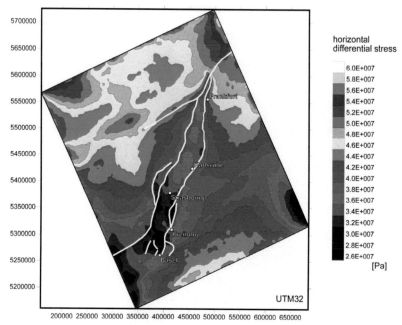

Figure 4.17: Calculated horizontal differential stress magnitudes (S_{Hdiff}) at 3000 m depth. Within the graben an increase in S_{Hdiff} magnitudes from the South to the North is observable.

Fracture Potential (FP; ECKERT & CONNOLLY, 2004; see section 2.1.4) is a parameter that allows the identification of areas of increased likelihood of brittle tectonic activity. It is used here as an analogue for deformation on second and third order faults (1000's to 100's m long) known to exist but not included in the regional scale URG model. Combining the analysis of FP results with other derivatives of the stress tensor such as Regime Stress Ratio (RSR) enables a prediction of the distribution of the expected style of deformation for the model area. An FP analysis was carried out to determine regions within the sedimentary infill of the graben that suggests an increased likelihood of tectonic activity. For this analysis a post-processing rheology with cohesion (c_0) of 5.6 MPa and a coefficient of internal friction (μ_i) of 0.8 was used. No tensile fracturing (tFP) is predicted by the model. Increased shear fracturing (sFP) is predicted to occur in two regions within the sedimentary fill of the graben (Figure 4.18). This implies that in these regions both shear fracture generation and reactivation of appropriately oriented faults are the active mechanisms of brittle deformation within the sedimentary graben infill. The modelled location of the two maxima of sFP predicted is in good agreement with the location of recent depocentres in the URG (Figure 4.18). The FP calculation based on the modelled stress state demonstrates the possible genesis of the recent depocentres under the boundary conditions applied to the model. In general, sFP is relatively decreased within the graben in the vicinity of appropriately oriented frictional contact surfaces implemented.

In order to analyse the influence of the frictional contact surfaces implemented on the kinematic loading of the URG system, elasto-plastic material properties are used (see Table 4.3). Plastic failure occurs within the model after the yield stress σ_y is exceeded. The magnitude of plastic strain in the vicinity of the frictional contact surfaces illustrates the relative degree of decoupling of the system. Relatively decreased magnitudes of plastic strain are limited to regions which exhibit a high degree of decoupling (Figure 4.19). Here, accumulative differential loading is dissipated by reactivation of pre-existing fault segments (i.e. the frictional contact surfaces). In contrast, relatively increased magnitudes of plastic strain indicate a higher degree of coupling over the fault segment. In the central segment of the URG the accumulative differential loading cannot be dissipated by the reactivation of pre-existing fault segments (i.e. implemented border faults) due to their unfavourable orientation (Figure 4.19). Here, increased magnitudes of plastic strain indicate that deformation occurs along second order fault structures within the graben not implemented in the regional scale URG model.

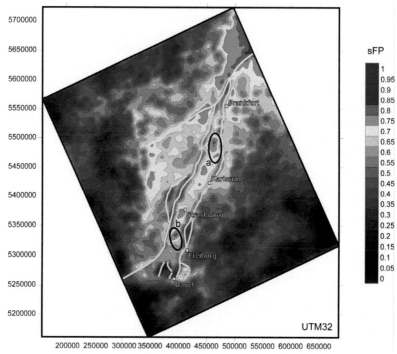

Figure 4.18: Calculated values of fracture potential at 1000 m depth. Within the URG two regions of shear fracture potential (sFP) that coincide with the location of the Quaternary depocentres (a: Heidelberger Loch; b: Geiswasser Basin; see Figure 1.4). In addition, increased sFP is indicated in the central segment of the URG to the west of Karlsruhe and to the South of Strasbourg.

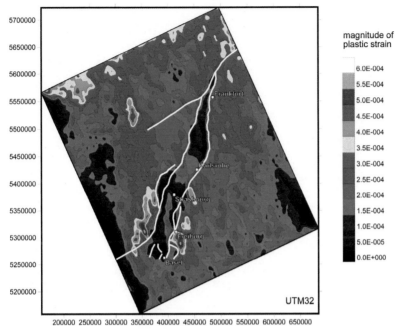

Figure 4.19: Calculated magnitudes of plastic strain at 3000 m depth. Within the graben the magnitude of plastic strain predicted by the model is increased for the central segment and for a region in the southern segment of the URG. In contrast, low magnitudes of plastic strain in the northern segment and in the vicinity of most fault segments implemented indicate the dissipation of accumulated differential loading due to slip along pre-existing faults (i.e. graben bounding faults).

4.5.4 Relative surface displacement and prediction of fault slip magnitudes

Relative surface uplift was induced during the modelling study in order to obtain sufficient relative displacement and fault slip magnitudes within the northern URG. This uplift was generated by uplift of the Moho induced by gravitational de-stressing of a mantle with initially higher density values applied during the gravitational pre-stressing step. This regional uplift pattern is induced sub-parallel to the axis of the URG (Figure 4.20). By applying this boundary condition, the highest surface uplift is induced in the south of the URG, in the vicinity of the Vosges Black Forest Dome. Here, an uplift equivalent of a rate of ~0.4 mm/a is induced. A minor locus of increased surface uplift is predicted for the area east of Heidelberg, corresponding to the Odenwald basement high and for the northern Vosges Mountains, north of the Saverne Basin. Areas of relative subsidence predicted by the model are the margin of the Paris Basin in the NW and the Molasse Basin in the SE (Figure 4.20).

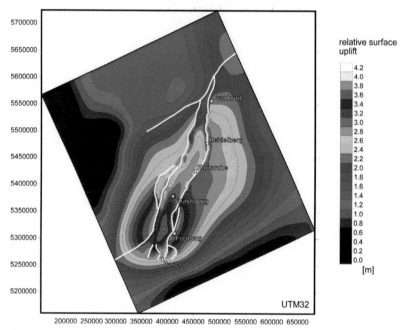

Figure 4.20: Induced magnitudes of of relative surface uplift. Along the graben axis the induced uplift is gradually decreasing from the South (i.e. Vosges Black Forest Dome) to the North, whereas the axis of highest uplift is slightly rotated clock-wise relative to the graben axis.

In order to evaluate the induced displacement pattern, modelled fault slip rates are compared to determined slip rates. All of the reference data was determined for the URG border faults (Table 4.5). The prediction of fault slip rates in the northern URG is in good agreement with the observed data. In the southern URG, however, the slip rates are significantly underestimated, when compared to levelling data. For this region, the applied boundary conditions are insufficient to simulate the kinematic behaviour of the border faults of the URG as revealed by geodetic observation. A reason for the systematic underestimation of fault slip rates in the southern URG might be, that the induced surface uplift of ~0.4 mm/a exceeds the actual uplift rates in this area.

Location (No. in Fig. 4.23)		Period [ka]	Method	Reference	Fault slip rate [mm/a]	Fault slip rate of model after 10 ka [mm/a]
Northern URG	Western Border Fault near Osthofen (1)	800	Terrace mapping	PETERS & VAN BALEN, 2007	0.06 – 0.13	0.03 – 0.07
	Western Border Fault near Osthofen (1)	19	Paleoseis-mological trenching	PETERS et al., 2005	0.03	0.03
	Eastern Border Fault at Heidelberger Loch (2)	1800	Sediment thickness	BARTZ, 1974	0.2	0.1
Southern URG	Weinstetten Fault near Freiburg (3)	0.025	Precise levelling	RÓZSA et al., 2005	0.17 max. 0.7	0.08
	Eastern Border Fault at Freiburg (4)	0.059	Precise levelling	RÓZSA et al., 2005	0.25 max. 0.45	0.01

Table 4.5: Fault slip rates of the URG border faults obtained from geomorphological, geological and geodetic investigations compared to modelling results of this study. For the location of the fault segments see Figure 4.21.

The more likely reason for this underestimation of fault slip magnitudes might be the simplified displacement boundary condition applied on the south-western border of the model. The applied homogeneous displacement at this boundary implies that no differential displacement can occur across the Rhine Saone Transfer Zone. This leads to a more constraint southern end of the URG and thus to an underestimation of crustal extension in this area.

It is important to note that the regional scale URG model can only address the relative surface uplift at long wavelengths as induced by the basal boundary condition, since local heterogeneities are not present in the model. Furthermore, the recent uplift history of the URG area is largely unconstrained and only relative uplift rates have been determined using geomorphological methods (e.g. PETERS & VAN BALEN, 2007; PETERS, 2007). Lastly, the amount of shoulder uplift inferred from geological data (ILLIES, 1974A; SCHWARZ, 2006; Eocene to present-day and Oligocene to present-day respectively) occurred over a time span longer than the one being modelled in this study. The tectonic behaviour of the URG changed significantly near the Oligocene – Miocene boundary (e.g. SCHUMACHER, 2002) and most of the uplift of the Vosges and Black Forest occurred during the middle to late Miocene (DEZES et al., 2004; ZIEGLER & DEZES, 2007). The sedimentary infill of the URG indicates increased tectonic activity and high subsidence within the northern URG from the Miocene (Figure 1.5). Fission-track data have been used to infer high uplift rates for the Vosges Mountains and Black Forest after this change in kinematic behaviour (LINK et al., 2004). In addition, the southern URG is characterised by subsidence from the late Pliocene onward (e.g. Ziegler & Dezes, 2007).

The present model underestimates the elevated uplift of the Rhenish Massif documented from the analysis of river terraces and river incision (MEYER & STETS, 1998; VAN BALEN et al., 2000). This uplift is regarded to as the isostatic response of the lithosphere to an asthenospheric convective instability, referred to as the Eifel Plume (GARCIA-CASTELLANOS et al., 2000, RITTER et al., 2001). Since no attempt was made to include this effect below the Rhenish Massif in the model presented herein, as this area is located at the margin of the model, the Rhenish Massif uplift predicted is significantly lower than that derived from geomorphological data (MEYER & STETS, 1998; VAN BALEN et al., 2000). Detailed information on the Pleistocene uplift of the Rhenish Massif resulting from the Eifel Plume was obtained from the thermo-mechanical modelling study by GARCIA-CASTELLANOS et al. (2000), which focused only on this region.

In detail, a disparity exists between the vertical displacement pattern obtained from the present model and the subsidence pattern within the URG (Figure 4.2.1). It is thought that this disparity occurs because of the geometric simplifications associated with only including the first order faults in the model. The dense network of second and third order intra-graben faults is not included, and hence the slip along them, which accommodates a large proportion of the actual uplift and subsidence, is not considered in the model. This means that deformation on structures below the model resolution cannot be directly determined. It is reassuring to note however that the model predicts high differential uplift between shoulder and graben areas along the graben bounding faults, as observed in nature (Figure 4.20).

In order to analyse the kinematic behaviour of the URG border faults, the magnitude of the gradient of the relative surface displacement is calculated. The gradient magnitude ($|\bar{g}|$) describes the change in magnitude of relative surface displacement at any point on a surface and is defined as:

Equation 4.4
$$|\bar{g}| = \sqrt{\left(\frac{\partial z}{\partial x}\right)^2 + \left(\frac{\partial z}{\partial y}\right)^2} \quad ,$$

where z is the surface displacement. The gradient of the relative vertical surface displacement at the position of a fault surface is a relative measure of the vertical displacement and for the models presented herein for the extensional slip component along that particular fault. Using the gradient of the relative vertical surface displacement enables fault segments with relatively increased vertical displacement to be identified (Figure 4.21).

The distribution of increased vertical displacements along the graben bounding faults shows two distinct maxima (Figure 4.21). In the northern segment of the URG most of the vertical displacement is accommodated at the Eastern Border Fault (EBF), whereas in southern segment of the URG most of the vertical displacement is accommodated by the Western Border Fault (WBF). In general, the model predicts increased vertical displacements along N-S and NNW-SSE striking faults.

Similarly to the gradient of relative vertical surface displacement, the gradient of the relative horizontal surface displacement at the position of a fault surface is a relative measure for the strike-slip component along that particular fault. Using the gradient of the relative horizontal surface displacement fault segments with relatively increased horizontal displacement can be identified (Figure 4.22).

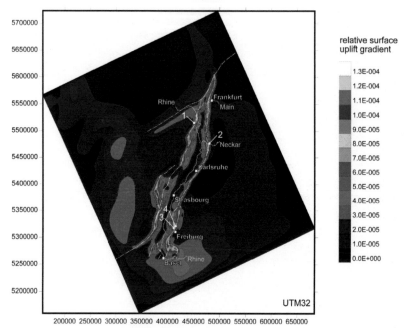

Figure 4.21: Calculated values of the relative surface uplift gradient. Fault segments characterised by an increased dip-slip component (i.e. extensional faulting) are indicated by an increased relative surface uplift gradient. Locations of documented fault slip data are indicated by numbers (1-4) as shown in Table 4.5.

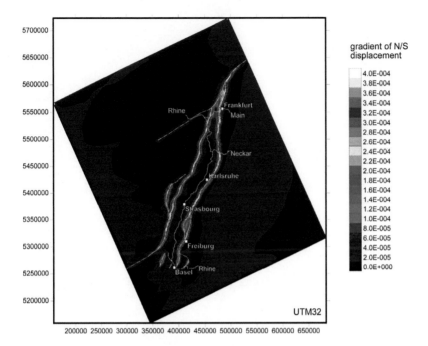

Figure 4.22: Calculated values of the N/S component of the relative surface displacement gradient. Fault segments characterised by an increased strike-slip component are indicated by a higher relative surface displacement gradient.

The distribution of increased horizontal displacements along the graben bounding faults is more uniform than the vertical component. Several local maxima are predicted along the EBF as well as along the WBF. In general, the model predicts increased horizontal displacements along NNE-SSW striking fault segments. For comparison, the vertical and horizontal displacement gradients have been normalised and combined in Figure 4.23.

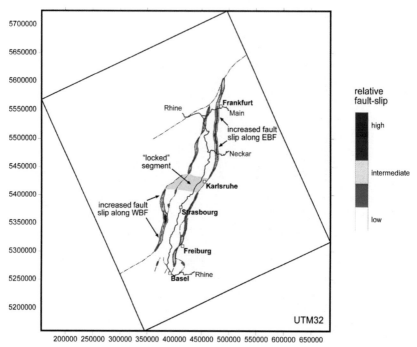

Figure 4.23: Relative fault slip derived from surface displacement gradients. The model predicts a change in asymmetry of the graben. In the northern segment of the URG most of the displacement is localised along the EBF, while in the southern segment most of the displacement occurs along the WBF. Note the relatively locked central segment of the graben.

The fault slip pattern predicted by the model is in good agreement with the compilation of graben bounding fault segments with documented Pleistocene activity (based on data by BREYER & DOHR,1967; AHONER & SCHNEIDER, 1974; ILLIES, 1975; CUSHING et al., 2000; HAIMBERGER et al., 2005; PETERS, 2007; Figure 4.24).

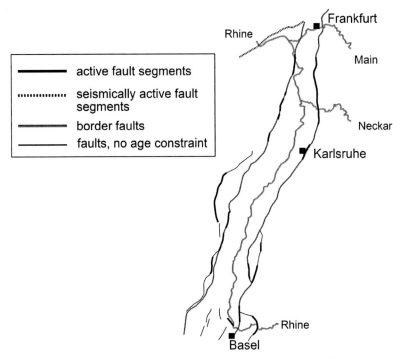

Figure 4.24: Compilation of segments of the graben bounding faults with documented Pleistocene activity. For references see Figure 1.10.

4.6 Discussion of modelling results and conclusions

At a regional scale, the state of in-situ stress predicted by the URG global model is one possible approximation of the state observed for the URG system. Within the limits of the method used and the uncertainties of the reference data the calibrated stress magnitudes are in good agreement with the available benchmarking data (see section 4.4.3). Furthermore, fault plane solution data indicate transtensional to extensional faulting as the dominant style of deformation within the upper crust of the URG area (e.g. DELOUIS et al., 1993; PLENEFISCH & BONJER, 1997; HINZEN, 2003). The analysis of the RSR value shows that this stress state could be reproduced by the model for the entire depth range of the upper crust (Figure 4.15). Therefore, the global model results are appropriate to subsequently load more complex local scale models to investigate the reactivation potential of second and third order fault structures in more detail (see chapter 5 and 6).

The modelling results imply a relative decrease of the horizontal stress components above the Vosges Black Forest Dome. In contrast, an increase of the horizontal stress components is predicted for the Rhenish Massif and the Molasse Basin. This variation in stress magnitudes is related to the topographies of the lithospheric layer boundaries (i.e. crustal thickness). As suggested by several previous authors (e.g. ILLIES, 1974A; ILLIES & GREINER, 1976, SCHUMACHER, 2002) the far-field load on the graben imposes a sinistral shear couple. The URG global model suggests that this sinistral shear couple is acting on a region typified by varying transtension and that fault displacement is occurring on a series of inherited structures, which will reactivate in response to their local loading (Figure 4.25). In this context, the model predicts a change in asymmetry of the graben. In the northern segment of the URG most of the displacement is localised along the EBF, while in the southern segment most of the displacement occurs along the WBF (Figure 4.27). This prediction is in good agreement with the observation of faults with documented Pleistocene activity (Figure 4.24). Relatively decreased fault slip is predicted for the central segment of the URG. This kinematic behaviour also becomes evident by analysing the magnitude of plastic strain predicted by the model (Figure 4.19). Plastic strain accumulates in the central segment of the URG, whereas low magnitudes of plastic strain in the vicinity of most fault segments indicate the dissipation of accumulated differential loading due to slip along pre-existing faults (Figures 4.19 and 4.25).

The 3D stress state predicted for the URG provides significant insight into its' recent kinematic behaviour. By using the FEM the influence of various topographies, geometric structures and boundary conditions on the present-day stress state and the recent kinematic behaviour of the URG are analysed. The boundary condition analysis of the URG model shows that the Moho topography serves as a critical loading condition on the regional kinematic model. The majority of upper crustal modelling studies assume a horizontal base of the model, which is only free to move in its plane (e.g. GÖLKE, 1996; DIRKZWAGER, 2002; LOPES CARDOZO, 2004; ECKERT, 2007). In contrast, this study clearly demonstrates that such an approach is inappropriate for the URG where, in addition to the far-field loads associated with the Alps and the Atlantic mid ocean ridge, vertical strain played an important role in the Quaternary graben history.

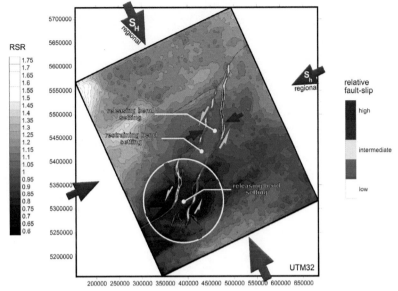

Figure 4.25: Kinematic model of the recent tectonic behaviour of the URG derived from modelling results. The URG acts as a sinistral shear couple in a region characterised by varying transtension induced by far-field loading. Additionally, the relative dominance of the horizontal stress components is a function of the crustal thickness. In the vicinity of the Vosges Black Forest Dome the model predicts a higher dominance of the vertical stress component, which is induced by the elevated surface topography in this region (region of white circle). The loading of the system is dependant on the orientation of the URG and its bounding faults relative to the imposed stress field. In the northern and southern segments of the URG appropriately oriented faults can accommodate for most of the induced deformation, whereas in the central segment the accumulative differential loading can not be dissipated by the reactivation of pre-existing fault segments.

PREDICTING THE IN-SITU STRESS STATE AND THE FAULT REACTIVATION POTENTIAL IN THE CENTRAL SEGMENT OF THE URG (SUB-MODEL A)

5.1 Introduction

Many previous authors argued that the far-field load imposes a sinistral shear couple on the URG (e.g. ILLIES, 1974A; ILLIES & GREINER, 1976, SCHUMACHER, 2002). The results of the large scale (global) modelling conducted in this study also suggests sinistral reactivation of the graben system and its bounding faults is plausible. Under these conditions, the central segment of the URG forms a restraining bend, since the strike of the graben bounding faults here changes towards the NE/SW when compared to the general NNE/SSW trend of the graben (Figure 4.27). Moreover, in the central segment the graben bounding faults are oriented nearly perpendicular to the maximum horizontal stress component of the present-day regional stress field (Figure 1.12). This means that the regionally imposed loads are less likely to be dissipated by movements along pre-existing fault structures due to their unfavourable orientation. This hypothesis is supported by the distribution of Quaternary sediments within the graben, which show a distinct thickness minimum in the central segment of the graben (Figure 1.4 B). Both, the northern and central segments of the URG are characterised by few and scattered earthquakes (BONJER et al., 1984; see section 1.4.3, Figure 1.13). In contrast to the northern segment, damaging earthquakes have occurred more frequently in historical times in the central segment (e.g. Rastatt 1933, Seltz, 1952; Figure 5.1). The majority of seismic events are thought to show predominantly horizontal (i.e. a high component of strike-slip) displacements, with seismicity caused by extensional faulting occurring less frequently (AHORNER & SCHNEIDER, 1974).

The main focus of the local scale modelling study of the central segment of the URG (*sub-model A*) is to investigate the reactivation potential of pre-existing faults under the present-day stress field. This may permit to identify fault segments that are prone to reactivation contemporary tectonic conditions. Furthermore, surface displacements and relative fault slip are investigated. The modelling results are compared with focal mechanisms determinations and known relative fault block movements. Additionally, the resolved slip tendency values are compared to the distribution of damaging earthquakes known from instrumental earthquake recordings.

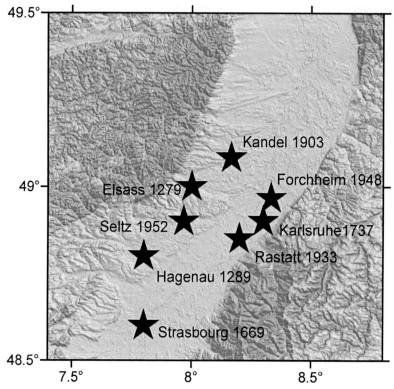

Figure 5.1: Concentration of historical earthquakes with intensities (I$_o$) VII and documented damages in the central URG (after LEYDECKER, 2005A; SCHWARZ et al., 2006).

5.2 Construction of sub-model A

The 3D Finite Element model (FEm) presented in this chapter comprises the central segment of the URG, parts of the Saverne Basin and the adjacent shoulder regions (Figure 5.2). The model contains two lithospheric layers: the lower crust and the upper crust consisting of shoulder and graben units and Cenozoic sedimentary fill of the graben (Figure 5.2). Due to the limited resolution of the model a uniform thickness of Cenozoic sediments of 2 km is assumed. Analogous to the construction of the global model, the geometric basis of the sub-model of the central segment of the URG (*sub-model A*) is an Earth model that includes fault surfaces, the surface topography and the topography of the brittle ductile transition zone (BDTZ). In contrast to the URG global model, the upper mantle is neglected. Instead, a planar basal boundary of the model is defined within the lower crust, since the effect of Moho uplift in the area of the URG is transferred as a boundary condition from the global model to the basal upper crustal nodes of the sub-model.

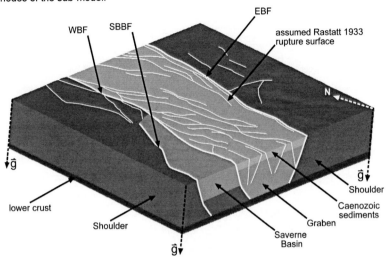

Figure 5.2: Oblique view of the spherical geometry of *sub-model A*. Gravitational acceleration is acting towards the centre of the Earth. The model dimensions are approximately 95 X 84 km, and the total thickness of the model is approximately 19 km. The model comprises two lithospheric layers, three different upper crustal units and a simplified fault model of first (bold white lines) and second (thin white lines) order fault surfaces implemented as frictional contact surfaces (see section 3.1.2; WBF, Western Border Fault; EBF, Eastern Border Fault; SBBF, Saverne Basin Border Fault). For references of the fault model used see Figure 1.9. At the model boundaries, the geometry of first order fault surfaces is identical to these of the global model to reduce stress concentrations at the interface between the two models.

Analogous to the global model, lateral depth dependent density variations are assigned to the lower and upper crustal units. The density variations are derived from Bouguer anomaly data and are assigned as a field dependent variable to the nodes of the model (see Figure 4.2). For the URG *sub-model A* first-order tetrahedral elements are used. Specifications on the element type and quantity of the various units are given in Table 5.1.

Model Unit	Element Type	~Element Size [m]	~Quantity
Cenozoic sediments	C3D4	500	302,000
Graben & Saverne Basin	C3D4	500-2000	904,000
Shoulders	C3D4	500-2000	591,000
Lower crust	C3D4	2000	15,000

Table 5.1: First-order tetrahedral elements are used for the discretisation of the various model units. A maximum resolution of 500 m is used within the graben and along the frictional contact surfaces implemented to represent the faults. Average element size and quantity are presented for each model unit.

A compilation of first- and second-order upper crustal faults are implemented in both the shoulder and graben units as frictional contact surfaces. For all surfaces a uniform coefficient of static sliding friction (μ_{static}) of 0.6 is assigned, analogous to the faults of the URG global model (see section 4.2). In order to minimise stress concentrators at the sub-model boundaries, identical listric geometries of the border faults are used. The boundary faults, as well as the second order fault structures implemented, mainly dip 60° with the dip direction inferred from 2D seismic sections. The maximum depth of the various fault surfaces implemented is inferred from two deep seismic profiles (Figure 1.8). The geometry of both, the graben bounding faults and second order fault surfaces are relatively well constrained for the shallower part of the profiles. For the construction of the sub-model it was assumed that first order fault structures terminate at the brittle ductile transition zone, whereas second order fault structures extend into the basement underneath the graben to a maximum depth of approximately 10 km (BRUN et al., 1991; WENZEL et al., 1991; Figure. 1.8). Within the central segment of the URG the graben exhibits an asymmetry in the faulting pattern. On the eastern border, deformation is more localised on the Eastern Border Fault and parallel to (NE/SW to NNE/SSW striking) second order structures, whereas on the western border of the graben the deformation is more distributed on predominantly N/S striking second order fault structures (Figures 1.9 I, 5.2). The fault model implemented in *sub-model A* is based on a compilation by Peters (2007; see Figure 1.9).

5.3 Modelling parameters and boundary conditions

During the sub-modelling analysis, the appropriate boundary nodes of the subsequent model are constrained by displacements interpolated from the global model. Using this technique, the simplified displacement field described in section 4.3 is transferred to the boundary nodes of the sub-model. Similar elastic properties and behaviour to those of the global model are used for the various units of the sub-model. In contrast to the global model, three upper crustal units are defined. Within the sub-model, the highly fractured graben interior and the less fractured shoulder regions as well as the sedimentary fill of the graben are honoured. A dense network of second and third order fault surfaces is implemented in the sub-model volume. No plastic material behaviour is used to simulate the deformation within the model volume. It is thus assumed that most of the deformation occurs along pre-existing fault surfaces and thus high stress accumulations necessary to initiate failure are less likely to occur (e.g. AHORNER & SCHNEIDER, 1974; BONJER et al., 1984; FRACASSI et al., 2005). The material properties of the units implemented are given in Table 5.2.

Model Unit	Rock Type	Density [kgm⁻³]	Young's Modulus [GPa]	Poisson's Ratio	Reference
Cenozoic sediments	Sediments	2450-2550	30	0.275	TURCOTTE & SCHUBERT, 2002
Graben & Saverne Basin	fractured basement	2450-2750	45	0.275	TURCOTTE & SCHUBERT, 2002
Shoulder	Sandstone, Granite and Gneiss	2550-2750	55	0.275	TURCOTTE & SCHUBERT, 2002
Lower crust	Anorthosite	2750-2850	83	0.4	TURCOTTE & SCHUBERT, 2002

Table 5.2: Elastic properties assigned to the 3D FEm units.

5.4 Loading procedure

Within the sub-modelling analysis, the sub-model is linked to the global model only through time dependent values of nodal variables determined during the global model analysis (ABAQUS™/Standard version 6.6-1 Analysis User's Manual, §7.3.1). The sub-model is solved in a separate analysis and hence no time dependent values of variables defined at the element integration points (e.g. the stress tensor) are transferred during this type of analysis. Because only nodal displacements can be transferred on subsequent models from the global model, the loading procedure of the sub-model has to be identical to the URG global model and comprises the following loading steps:

- a gravitational pre-stressing step to establish realistic depth related stress magnitudes within the model and to counteract excess elastic compaction,

- a tectonic pre-stressing step to simulate the effect of the long-term deformation history of the URG on the present-day state of stress, and

- an analysis of the short-term (10 ka) deformation pattern of the central segment of the URG induced by lateral loading of the system.

5.4.1 Gravitational pre-stressing model

In the first step the model is only loaded by gravitational acceleration towards the Earth's centre and is basally and laterally constrained by a displacement field interpolated from the gravitational pre-stress step of the URG global model (Figure 5.3; see section 4.4.1).

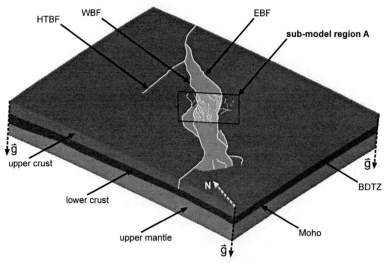

Figure 5.3: Gravitational pre-stressing of *sub-model A*. The model is loaded only by body forces induced by gravitational acceleration towards the Earth's centre. In order to achieve a model with relatively increasing horizontal stresses towards the Earth's surface, the model is basally and laterally constrained by nodal displacements interpolated from the pre-stressing step of the URG global model. WBF, Western Border Fault; EBF, Eastern Border Fault; HTBF, Hunsrück Taunus Border Fault. Note the increase in resolution due to the implementation of 2nd and 3rd structures relative to the global model.

5.4.2 Tectonic pre-stressing model

Similar to the global model analysis, the stress tensor obtained after the gravitational pre-stressing step is used as the initial stress condition for the tectonic pre-stressing step. This gravitational loaded model is then laterally deformed by transferring nodal displacements from the calibrated and benchmarked tectonic pre-stressing step of the global model on the appropriate basal and lateral boundary nodes of the sub-model to obtain a reasonable approximation of the state of stress for the central segment of the URG (Figure 5.4).

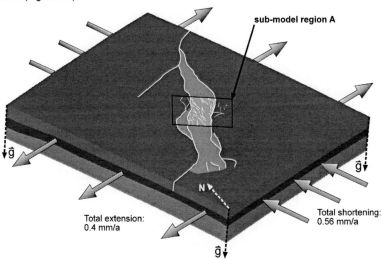

Figure 5.4: Oblique view of the tectonic loading of *sub-model A*. Arrows indicating the simplified lateral loading conditions applied perpendicular to the global model boundaries. Nodal displacements obtained during the global model analysis are transferred to the appropriate basal and lateral nodes of the sub-model region.

5.4.3 Modelling the short-term deformation pattern of the central URG

In order to address the short term surface displacement pattern of the central URG area, the boundary conditions obtained from the global model for a time increment of 10 ka are transferred to the *sub-model A* boundaries. The modelled vertical surface displacements are compared to uplift data derived from geological data (e.g. sediment distribution, relative uplift of intra-graben fault blocks). In addition to the surface uplift predictions, modelled relative fault slip rates are compared to available fault slip data.

5.4.4 Modelling the static stress transfer (ΔCFS) induced by the 1933 Rastatt earthquake

In order to analyse the static stress transfer induced by the 1933 Rastatt earthquake, the stress tensor generated by the tectonically loaded model is used as the initial stress state. At the time increment prior to the earthquake event, no relative fault movement is assumed for the entire system. At the time of the earthquake event, only relative slip along the implemented rupture plane within the brittle upper crust is considered. Using the concept of ΔCFS, it is investigated whether fault planes in the vicinity of the earthquake event are brought closer to shear failure (see section 2.2.3). Additionally, the results of the ΔCFS analysis are compared to the distribution of slip tendency along the modelled fault surfaces.

5.5 Modelling results

The purpose of *sub-model A* is to obtain an approximation of the present-day state of stress affecting the central segment of the URG. In order to do this, the influence of the dense network of second and third order fault surfaces on the state of stress, the kinematic behaviour and the displacement pattern must be considered. In this section the modelling results for the central segment of the URG are presented.

5.5.1 Prediction of the state of stress in the central segment of the URG

The state of stress predicted by *sub-model A* is similar to the prediction of the global model (compare section 4.5.2). Identical rheological properties are used except for the upper two kilometres of sedimentary fill of the graben. Furthermore, the complete loading procedure of the global model is transferred on the sub-model boundaries using the ABAQUS™ *submodelling* technique. For comparison, the modelling results are presented at a uniform reference depth of 3000 m below sea level.

A transtensional state of stress is predicted for the central segment of the URG (Figure 5.5a). Variations in the transtensional state of stress are induced by spatially varying material properties, crustal thickness, surface topographies and differential loading of the faults. The sub-model predicts a similar RSR magnitude distribution to the global model (Figure 4.15). Within the graben, the RSR magnitude is relatively increased in the sub-model due to a relative decrease in S_v magnitudes predicted as a result of the lower densities of the sedimentary fill of the URG and the lower surface topography (RSR values of 0.9 to 1.2). In general, lower RSR magnitudes are predicted for the shoulder regions (RSR values of 0.6 to 1.0). On the western shoulder RSR magnitudes are relatively decreased in the vicinity of the WBF due to the dissipation of horizontal stresses induced by fault slip along the WBF and second order fault structures. On the eastern shoulder, the influence of the surface topography on the state of stress predicted is very clear. The magnitude of RSR is relatively increased in the region of the Kraichgau Trough in the North, whereas in the region of the northern Black Forest in the Southeast the model predicts a stronger dominance of the vertical stress component and a relative decrease of the maximum horizontal stress component S_H, which results in lower RSR values.

In order to qualify the relative likelihood of brittle failure predicted by *sub-model A* relative magnitudes of the second invariant of the stress tensor (i.e. von Mises stress, σ_M) are analysed. The sub-model results show a similar σ_M magnitude distribution compared to the global model (Figure 4.16). A relative increase in magnitudes of σ_M is predicted for the shoulder regions, whereas relatively decreased σ_M magnitudes are predicted for the graben region (Figure 5.5b). For the graben region a lower elasticity is defined resulting in a more isotropic state of stress. The distribution of relative σ_M magnitudes for the shoulder regions is primarily a result of surface topography. Relatively elevated σ_M magnitudes are observed in the model results for areas characterised by elevated surface topography (e.g. northern Black Forest south of Rastatt), whereas relatively decreased σ_M magnitudes are predicted for regions characterised by low surface topography (e.g. Kraichgau Trough east of Karlsruhe and graben region). Within the graben σ_M is relatively decreased in the vicinity of appropriately oriented frictional contact surfaces in the NE and SW of the model. A relative increase in modelled σ_M magnitudes is observable in the central part of the model in the North of the city of Rastatt.

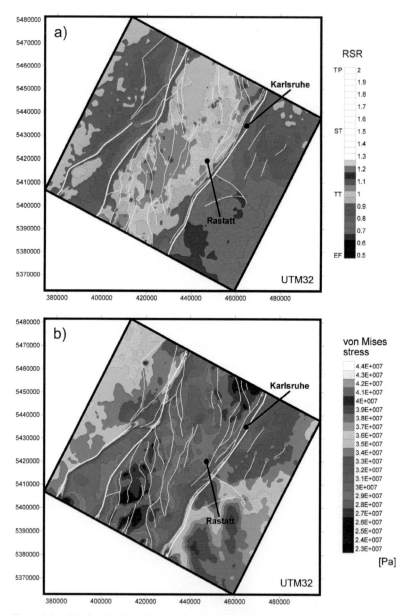

Figure 5.5a: Calculated value of the regime stress ratio (RSR; see section 2.1.2) at 3000 m depth. In general, the model predicts a transtensional state of stress for the central segment of the URG. RSR magnitudes are elevated within the graben relative to the shoulder regions. First

order fault structures also implemented in the URG global model are indicated by bold white lines. Thin white lines indicate second and third order fault structures only defined in the sub-model volume. Figure 5.5b: Calculated von Mises stress magnitudes (σ_M; see section 2.1.2.3) at 3000 m depth. On the scale of *sub-model A* σ_M shows a dependency on the surface topography for the shoulder region. σ_M is elevated in the region of the northern Black Forest in the SE of the model, whereas a decrease of σ_M magnitudes is visible for the region of the Kraichgau Trough in the NE of the model. Within the graben σ_M magnitudes are relatively decreased compared to the shoulder regions, whereas local variations are visible. Elevated σ_M magnitudes are predicted for the central part of the model. Note the locally induced stress concentrators at the intersection of the first order fault structures with the lateral model boundaries. Using a similar model geometry with the ABAQUS ™ *submodelling* technique, these stress concentrators are unavoidable. Fortunately, using an appropriate resolution the influence of the stress concentrators is only significant locally.

5.5.2 Prediction of relative surface displacement and fault slip magnitudes (sub-model A)

In order to analyse the induced deformation along the upper crustal faults within the central segment of the URG, displacements obtained from the global model for a time span of 10 ka are applied as nodal constraints on the appropriate boundaries of *sub-model A*. The sub-model predicts decreased magnitudes in relative vertical displacement due to the higher fragmentation of the brittle upper crust in response to the implementation of second and third order fault structures. The distribution of the relative uplift is similar to the results of the global model (Figure 4.20). The modelled uplift pattern suggests that increased relative uplift is induced along the eastern shoulder for both the northern Black Forest and the Kraichgau Trough. Within the graben, the model predicts a complex displacement pattern with varying uplift and subsidence of individual fault blocks (Figure 5.6a). At a local scale the model shows varying fault kinematics within the overall transtensional state of stress predicted. The vertical surface displacement modelled indicates that both extensional faulting as well as inversion of extensional faults occurs within the upper kilometres of the crust. It is seen that NNE/SSW oriented faults are more likely to be inverted than N/S oriented faults due to their unfavourable orientation relative to the imposed present-day stress field. The inversion of intra-graben faults is also indicated by earthquake focal mechanism solutions for the URG. AHORNER (1974) describes the dominance of the horizontal strain direction for most of the earthquakes associated with the URG. Additionally, both extensional and compressional strain components are documented from earthquake focal mechanisms within the URG (e.g. Aʜᴏʀɴᴇʀ, 1974; Bᴏɴᴊᴇʀ et al., 1984; Pʟᴇɴᴇꜰɪsᴄʜ & Bᴏɴᴊᴇʀ, 1997). In comparison to the global model results, *sub-model A* predicts a more realistic vertical displacement pattern for the intra-graben region with varying extension and inversion along the second and third order intra-graben faults. The modelling results indicate that the prediction of relative surface displacements is primarily dependant on the geometric complexity of the model.

In order to visualise the kinematic behaviour of the fault surfaces, the magnitude of the gradient of the relative surface displacement is calculated using Equation 4.4 (Figure 5.6b). The distribution of increased vertical displacements along the graben bounding faults shows several maxima (Figure 5.6b). A distinct maximum of relative vertical displacement is predicted for a segment of the Eastern Border Fault (EBF) south of the city of Rastatt. This segment is characterised by a change in strike from the overall NE/ SW trend to a NNE/SSW trend. For the EBF system in general, the model predicts relatively decreased values of the relative surface uplift gradient. Similar to the EBF system, the Western Border Fault system shows a dependency of the relative uplift gradient on the orientation of the particular fault segment. Within the graben, the model also predicts that NE/SW oriented fault structures are less likely to accommodate relative vertical displacements.

Figure 5.6a (next page): Calculated values of relative surface uplift induced. The predicted surface uplift is highest for the eastern shoulder region for both the northern Black Forest and the Kraichgau Trough. Within the graben a complex vertical surface displacement pattern with varying relative uplift and subsidence is predicted. Note the inversion of extensional faults parallel to the Eastern Border Fault east of the city of Rastatt. Figure 5.6b: Calculated values of the relative surface uplift gradient. Fault segments characterised by an increased dip-slip component (i.e. extensional faulting) are indicated by an elevated relative surface uplift gradient.

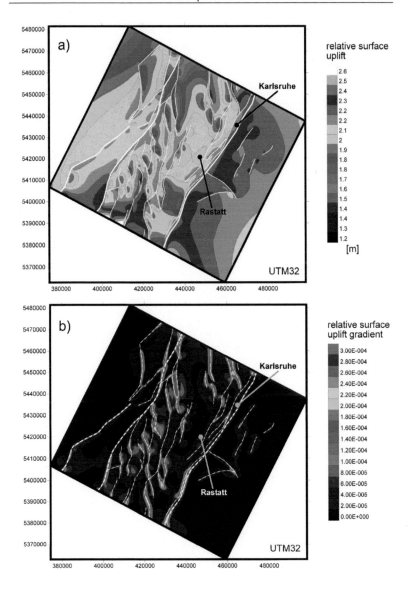

The gradient of the relative horizontal surface displacement at the position of a fault surface is a relative measure for the strike-slip component along that particular fault. Using the gradient of the relative horizontal surface displacement, fault segments with relatively increased horizontal displacement can be identified (Figure 5.7a). Similar to the predictions of the global model, the distribution of increased horizontal displacements is more uniform than the vertical component. The relative horizontal displacement gradient is less dependant on the orientation of each particular fault segment. Similar to the predictions of the relative vertical displacement gradient, the gradient of relative horizontal displacement shows a distinct maximum at a segment of the EBF south of the city of Rastatt. In order to visualise the total relative fault slip predicted by *sub-model A*, the vertical and horizontal displacement gradients have been normalised and combined in Figure 5.7b.

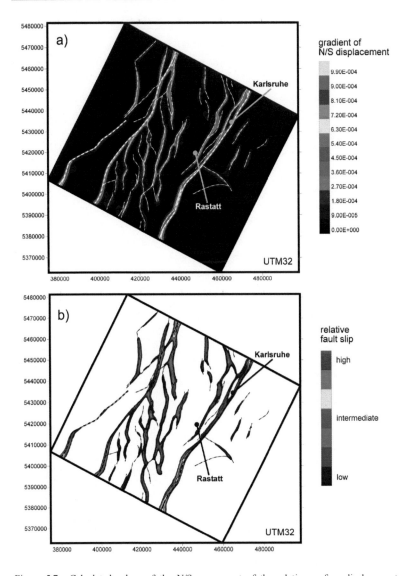

Figure 5.7a: Calculated values of the N/S component of the relative surface displacement gradient. Fault segments characterised by an increased strike-slip component are indicated by an elevated relative surface displacement gradient. Figure 5.7b: Relative fault slip derived from surface displacement gradients.

5.5.3 Dilation and slip tendency analysis for sub-model A

In this section, the dilation tendency (DT; likelihood of tensile reactivation) and slip tendency results (ST; likelihood of shear reactivation) for upper crustal faults of the *sub-model A* domain are presented and discussed. The theory of the DT and ST parameters is described in section 2.3. Both methods are applicable to planar discontinuities (e.g. fault or fracture surfaces; MORRIS et al., 1996; FERRILL et al., 1999). In order to calculate the DT and ST parameters, a three dimensional fracture or fault model as well as the three dimensional approximation of the stress field are required. Therefore, stress results obtained by a three dimensional finite element analysis (FEA) are a useful basis to use for investigating risking parameters related to planar discontinuities.

Using FEA, two principle approaches exist to determine DT and ST values for three dimensional planar discontinuities. The first approach assumes that during the loading procedure or time span of consideration, slip along discrete fault surfaces is unlikely. Using this approach, frictional contact surfaces within the fe-mesh can be neglected or, in case of a sediment filled graben structure, represented as internal boundaries between sub-domains characterised by different material properties. An ST analysis of the entire URG using this approach was conducted by PETERS (2007). The DT and ST analysis presented in this section follows the second approach, in which a dense network of fault structures is implemented within the fe-mesh to give a so-called fault block model. In this case, loading of the contact surfaces (i.e. faults) during the analysis can result in displacement along these fault structures resulting in local stress perturbations and dissipation of stresses induced by the fault slip.

For the calculation of both risking parameters (DT and ST), the scalar and planar components and derivatives of the stress tensor (σ_1, σ_3, σ_n and τ) are obtained for each individual surface element of a particular fault surface using the post processing software GeoMoVie. After the calculation, DT and ST values are visualised on an identical fault model using goCad. The calculation of the DT and ST values presented in this study is based on the following assumptions:

- all fault surfaces are considered to be cohesionless (C_0 = 0) and have not regained cohesive strength due to cementation,
- all fault surfaces are considered to have the same uniform coefficient of static friction (μ_{static}).

- continuous fault slip is possible during the loading of the system. Note that this mimics creep, which is the typical mechanism of dislocation along appropriately oriented pre-existing fault surfaces within the URG.

In contrast to the DT parameter, which is determined independently from any physical properties of the surface except for its orientation relative to the imposed stress field, the ST results are additionally dependant on the coefficient of static sliding friction (μ_{static}). Unfortunately, friction laws are only phenomenological descriptions of the friction behaviour and are based on empirical studies (see section 2.3.1). Typical friction values for crustal faults are still a matter of debate (e.g. ZOBACK et al., 1987; SCHOLZ, 2000). A low value of μ_{static} leads to increased ST values (i.e. brings the fault closer to shear reactivation), whereas a higher value of μ_{static} decreases the calculated ST values and makes shear reactivation less likely.

Since no constraint on the absolute or relative coefficient of sliding friction for upper crustal faults within the URG system exists, a uniform value of μ_{static} of 0.4 is used for the ST calculations. This relatively low value of μ_{static} is used since it yields values of ST in a reasonable range for the URG. Additionally, it is optimal for visualisation. The ST values obtained in this study maybe underestimated because of the assumptions the calculation is based upon. The model considers continuous slip to be possible during the imposed loading of the system. This means that the ST values represent the likelihood of repeated slip on implemented fault surfaces after slip has occurred. In addition, the effect of pore-pressure on the slip tendency, which increases the likelihood of slip, is not directly considered in this modelling study (see section 2.3.2). Despite the limitations of both the ST and DT method, the combined analysis of dilation and slip tendency values can provide useful insight on the relative likelihood of fault reactivation within a geological system. The ST method can be used to identify fault segments, which are more likely to be reactivated due to their more favourable orientation. The relative likelihood of tensile reactivation of pre-existing fracture sets can be investigated using the DT method. Using the combined methods of DT and ST, results in relative estimates of the likelihood of shear and tensile failure. The relative risk between shear and tensile failure cannot be determined, implying that the slip tendency values cannot be compared directly with the dilation tendency values. Both methods are used here to identify fault segments and fault orientations for which the risking parameter indicates a higher potential for reactivation.

Figure 5.8a shows the calculated values of the dilation tendency for the *sub-model A* domain. DT values in the range of 0 ($\sigma_n = \sigma_1$) to 1 ($\sigma_n = \sigma_3$) are predicted for the fault model of this central segment of the URG. Relatively increased values of DT are predicted for the vertical, NNE/SSE and NW/SE trending fault structures in the eastern shoulder region due to their orientation relative to the transtensional state of stress. Furthermore, extensional fault structures throughout the model volume with NNW/SSE to NNE/SSW oriented strikes show increased DT values. In the area of Rastatt, the Eastern Border Fault (EBF) as well as a set of sub-parallel fault structures shows relatively decreased values of dilation tendency. Figure 5.8b shows the calculated values of slip tendency for the *sub-model A* domain assuming a static coefficient of friction of μ_{static} = 0.4. In general, ST is increased for NNW/SSE to NNE/SSW oriented extensional faults throughout the model volume.

Figure 5.8a (next page): Calculated magnitude of dilation-tendency (DT) for upper crustal faults within the central segment of the URG at a depth of 4 to 8 km. Figure 5.8b: Calculated magnitude of slip-tendency (ST) for upper crustal faults within the central segment of the URG at a depth of 4 to 8 km, assuming a static friction coefficient of μ_{static} = 0.4.

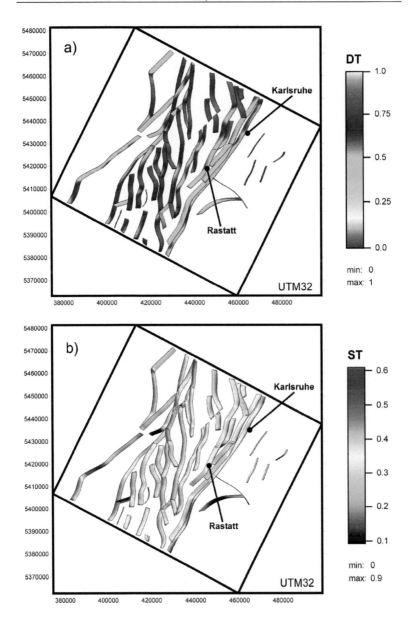

Analogous to the dilation tendency, slip tendency is relatively decreased for segments of the graben bounding faults. Similar to the DT results, the same segment of the EBF and the sub-parallel set of second order fault structures in the vicinity of Rastatt show decreased magnitudes of ST. In contrast to the DT results, vertical faults in general show relatively decreased ST magnitudes due to their unfavourable orientation relative to the transtensional state of stress. Within the central segment of the URG, relatively increased values of both DT and ST are predicted for N/S striking extensional fault structures. Due to the limitation of the method, it cannot be determined whether tensile or shear is failure is the primary mode of reactivation of upper crustal faults within the URG area. The results of the combined DT and ST analysis indicate that the majority of extensional fault structures that were established during the Eocene/Oligocene extensional phase of the URG are favourably oriented relative to the present-day stress field for both tensile and shear reactivation. Furthermore, the model indicates a high dilation tendency for the frequently occurring N/S orientated hydraulic fracture set (e.g. at the Soultz sous Forêts geothermal site; see section 4.4.3). Figure 5.9 relates the DT and ST values obtained to the distribution of fault strikes for the *sub-model A* domain.

Figure 5.9 a1) (next page): Rose diagram of fault segment strike orientations for upper crustal faults of the central segment of the URG compared to the distribution of predicted DT values. Increased DT values (≥ 0.5) are predicted for faults with strikes ranging from N340° to N10°. a2) Histogram indicating the distribution of DT values for the individual fault surface elements for *sub-model A*. The histogram shows a uniform distribution of DT values in the range between 0.45 and 0.85. The mean value of DT is 0.61. b1) Rose diagram of fault strike orientations for upper crustal faults compared to the distribution of predicted DT values. Relatively increased ST values (≥ 0.3) are predicted for faults with strikes ranging from N340° to N20°. b2) Histogram indicating the distribution of ST values for the individual fault surface elements. The histogram shows a distinct maximum at 0.36 for the ST distribution. The mean value of ST is 0.34. In general the model predicts higher dilation tendency values for the fault structures considered.

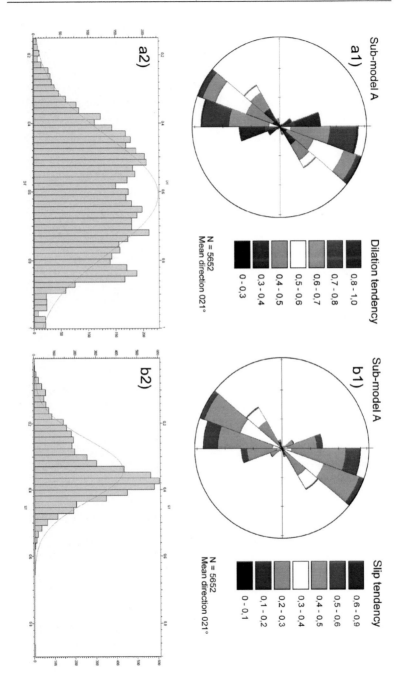

5.5.4 Static stress transfer (ΔCFS) induced by the 1933 Rastatt earthquake

The 1933 Rastatt earthquake is one of the strongest earthquakes that occurred in historical times within the URG system (Figure 1.14). The faulting mechanism of this earthquake event was interpreted to be predominantly strike-slip (HILLER, 1934; SCHNEIDER et al., 1966; AHORNER, 1967; AHORNER & SCHNEIDER, 1974). AHORNER & SCHNEIDER (1974) provide a combined focal mechanism solution of the 1933 event with a much weaker tectonic event that occurred in 1971 in the same region (Figure 5.10).

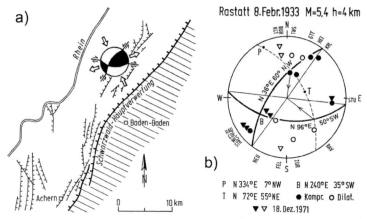

Figure 5.10: Combined focal mechanism solution for the 1933 and 1971 Rastatt earthquakes after AHORNER & SCHNEIDER (1974). a) Sketch map after BREYER & DOHR (1967) indicates the tectonic situation in the region of the two earthquakes. b) Compressional and dilatational stations for the combined focal mechanism solution of the two Rastatt events after AHORNER & SCHNEIDER (1974).

Following the interpretation of AHORNER AND SCHNEIDER the 1971 event had a comparable faulting mechanism. For the two Rastatt events AHORNER & SCHNEIDER (1974) identified a predominantly strike-slip mechanism with a thrust component. The two possible fault planes have a strike of SW/NE (sinistral; dip 60° towards NW) and WNW/ESE (dextral; dip 50° towards S). In the region of the Rastatt events, a graben parallel fault structure was identified by hydrocarbon exploration (Figure 5.10). Therefore, the SW/NE striking fault plane is favoured by most authors to be the rupture plane of the 1933 Rastatt event (e.g. HILLER 1934, AHORNER & SCHNEIDER, 1974). In contrast, FRACASSI et al. (2005) suggested an E/W striking rupture plane for this event.

Considering the results of the slip tendency analysis of this study and the results of PETERS (2007), the two identified possible rupture planes are characterised by low slip tendency values in the present day stress field. Other than the structural evidence, no independent method exists for identifying the actual rupture plane. Therefore, following the majority of authors, the SW/NE oriented graben parallel fault is used to simulate the 1933 Rastatt event in this study (Figure 5.2). The combined focal mechanism solution revealed a hypocentral depth of approximately 4 km (AHORNER & SCHNEIDER, 1974). In order to determine the possible rupture area of the 1933 Rastatt earthquake, a local magnitude (M_L) of 5.4 and a focal depth of 4 km is assumed (AHORNER & SCHNEIDER, 1974). To obtain the possible rupture area the local magnitude is converted to the moment magnitude (M_W) using the empirical relationship of GRÜNTHAL & WAHLSTRÖM (2003). WELLS & COPPERSMITH (1994) relating the moment magnitude (M_W) to rupture geometry parameter for different earthquake mechanisms using a regression analysis. Using the WELLS & COPPERSMITH (1994) model, the geometric properties of the modelled rupture plane have been obtained (Table 5.3). In order to model the possible rupture plane of the Rastatt 1933 earthquake a coefficient of static friction of 0.6 is used (see section 3.1.2).

hypocentral depth	height	length	Area
4 km	2.9 km	4.5 km	13 km^2

Table 5.3: Geometric properties of the rupture plane modelled. The hypocentre is assumed to be at the centre of the rupture plane. The average dip of the rupture plane is considered to be 65° and the dip direction is considered to be 316°.

In order to analyse the static stress transfer induced by the 1933 Rastatt earthquake, the stress tensor from the tectonically loaded model is used. Within the earthquake model, slip is allowed to occur only on the predefined rupture plane (Table 5.3). The concept of ΔCFS is used to investigate whether fault planes in the vicinity of the earthquake event are transferred closer to failure (see section 2.2.3). Additionally, the results of the ΔCFS analysis are compared to ΔST results (change in slip tendency due to the tectonic event) and the distribution of slip tendency modelled along the modelled fault surfaces. The calculation of the ΔCFS values presented in this study is based on the following assumptions:

- At the time increment prior to the earthquake event, no relative fault movement is assumed for the entire system.

- At the time of the earthquake event, only relative slip along the implemented rupture plane within the brittle upper crust is considered.

- During the tectonic event no change in pore fluid pressure (P_f) occurs.

- The coefficient of static friction is assumed to be 0.6.

In order to assess the model quality, the displacement on the fault surface is used to calculate the moment magnitude of the modelled earthquake. Using the approach of KANAMORI & ANDERSON (1975) the seismic moment (M_0) can be derived from the shear modulus (G), obtained from the elastic parameters defined within the model, the rupture area (A) and the average fault slip (D):

Equation 5.1 $$M_0 = G_{[dyne/cm^2]} A_{[cm^2]} D_{[cm]} \;.$$

The seismic moment (M_0) can then be used to determine the moment magnitude (M_W; e.g. STEIN & WYSESSION, 2003):

Equation 5.2 $$M_W = \frac{\log(M_0)}{1.5} - 10.73 \;.$$

Table 5.4 compares the moment magnitude derived from the modelled fault slip to the moment magnitude derived from the local magnitude (M_L) given by AHORNER & SCHNEIDER (1974).

M_W derived from model input data (M_L)	M_W derived from model results (D)
5.035	5.05

Table 5.4: Moment magnitude derived from the modelled fault slip compared to the moment magnitude derived from the available input data.

Figure 5.11 illustrates the construction and benchmarking procedure of the 1933 Rastatt earthquake model.

Input data

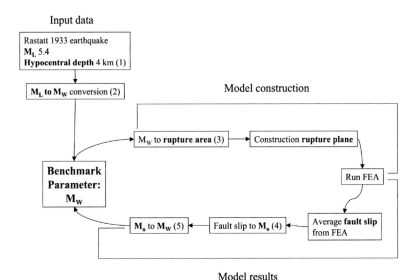

Figure 5.11: Diagram illustrating the construction and benchmarking procedure of the 1933 Rastatt earthquake model. References for the conversion steps are given by numbers: (1) AHORNER & SCHNEIDER, 1974; (2) GRÜNTHAL & WAHLSTRÖM, 2003; (3) WELLS & COPPERSMITH, 1994; (4) KANAMORI & ANDERSON, 1975; (5) STEIN & WYSESSION 2003.

The displacement modelled along the fault surface is also used to calculate a pseudo fault plane solution. For this purpose a Numerical Dynamic Analysis, NDA, (e.g. SPANG, 1972; SPERNER & RATSCHENBACHER, 1994) is conducted. Using the NDA, the mean orientation and R-ratio of the principal stress components can be derived from the displacement vectors measured on a fault plane. Figure 5.12 illustrates the results of the NDA for the simulated earthquake event. Since both the orientation of the fault plane and the orientation of the stress field are known from the modelling results, a realistic angle ($\theta = 75°$) between the reactivated rupture plane and the maximum principal stress component is used to calculate the pseudo fault plane solution. The results of the NDA reveal a strike-slip faulting mechanism with a minor extensional component for the modelled earthquake.

Sub-model A

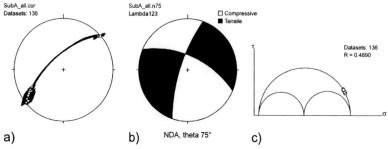

Figure 5.12: a) Three dimensional distribution of the slip vector on the rupture plane, fault lineation data plotted on lower hemisphere, equal area projection. b) Pseudo fault plane solution derived from the modelled slip vectors for the Rastatt event using NDA ($\theta = 75°$). c) Calculated three dimensional Mohr circle for the fault slip data.

The benchmarked model was then used to investigate the effects of the 1933 Rastatt earthquake on the stress field and possible earthquake triggering on adjacent fault surfaces. Figure 5.13 illustrates the induced surface displacement and the induced change in maximum shear stress. The surface displacement shows a distinct asymmetry with the greater magnitudes occurring in the structural hanging wall (i.e. graben block). The induced surface displacement is a function of the faulting mechanism and the structural composition of the affected volume. In contrast, the pattern of the change in maximum shear stress magnitudes is more symmetric indicating the almost pure strike-slip mechanism of the modelled earthquake (compare Figure 5.12).

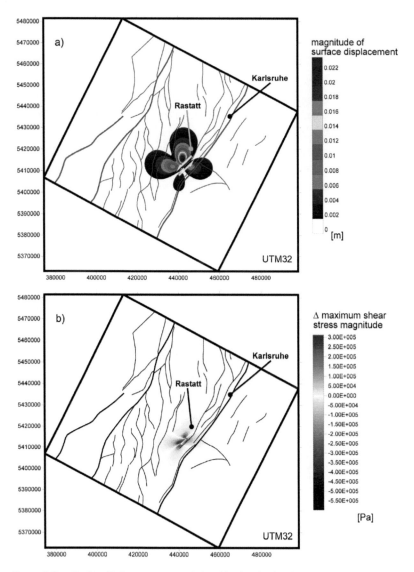

Figure 5.13 a: Surface displacement pattern induced by the simulated Rastatt 1933 earthquake. Note the asymmetry of the surface displacement pattern. Figure 5.13 b: Change in the maximum shear stress magnitude at a depth of 4 km induced by the simulated Rastatt 1933 earthquake. The transparent yellow surface indicates the rupture plane modelled. Contour lines represent a change in maximum shear stress magnitudes of 100,000 Pa.

Contrary to the commonly used ΔCFS (change in Coulomb Failure Stress) parameter, the change in the maximum shear stress magnitude (τ_{max}) is also used to study the volumetric change of the shear stress field induced by the modelled earthquake event (Figure 5.13 b). τ_{max} is a scalar value and can be calculated for each data point. In contrast, CFS is a parameter which depends on the orientation of a pre-existing fault surface relative to the stress field. Therefore ΔCFS can only be used as a scalar value to illustrate the volumetric stress change when an infinite set of parallel fault surfaces with known orientation is assumed throughout the model volume. In this study, ΔCFS is only calculated on the fault surfaces used to define the *sub-model A* geometry and the absolute change in shear stress is considered (see section 2.3.3; Figure 5.2; Figure 5.14 b).

Figure 5.14 (next page): Change in planar parameters induced on the fault model by the simulated Rastatt 1933 earthquake shown over a depth of 4 to 8 km. a: Change in slip tendency (ST) induced by the simulated Rastatt earthquake. The insert is showing the vicinity of the modelled rupture plane in more detail. b: Change in Coulomb Failure Stress (CFS) induced by the simulated Rastatt 1933 earthquake. The insert is showing the vicinity of the modelled rupture plane in more detail; A indicates a segment of the EBF system with reduced CFS, whereas the CFS is significantly increased for the fault segments marked with B.

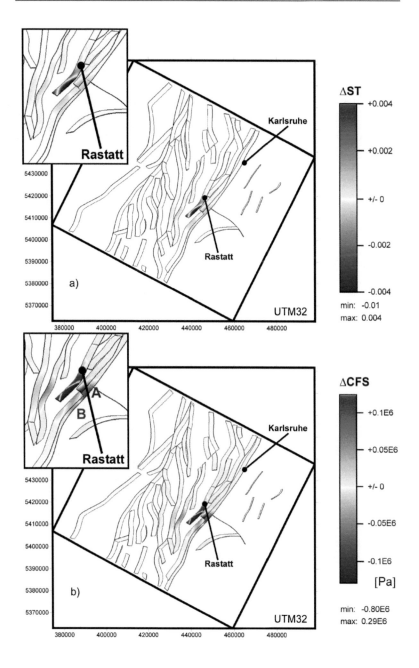

In addition to ΔCFS, the change in slip tendency (ΔST) is calculated for the surfaces of the considered fault model (Figure 5.14 a). The ΔST values show the change in the relative likelihood of fault slip dependant on the orientation of the fault surface relative to the stress field. The slip tendency (ST) is reduced significantly in the vicinity of the rupture surface. Adjacent to the rupture surface ST is increased on the same fault surface (Figure 5.14 a insert). The ΔST analysis reveals only a minor influence of the Rastatt 1933 earthquake on the state of stress along adjacent fault surfaces within the central segment of the URG. In contrast to the relative approach of ΔST, the ΔCFS values show the absolute change in Coulomb Failure Stress (CFS) magnitudes for adjacent fault surfaces induced by the slip along the rupture surface. CFS is reduced by up to ~0.8 MPa in the vicinity of the rupture surface (Figure 5.14 b insert). Additionally, CFS is reduced significantly for a segment of the EBF and a sub-parallel fault structure to the East of the modelled event (marked with A in Figure 5.14 b). Adjacent to the rupture surface CFS is increased on the same fault surface by up to ~0.3 MPa. Furthermore, CFS is increased for a segment of the EBF and a sub-parallel fault structure situated in the SE of the modelled event by up to 50 KPa (marked with B in Figure 5.14 b). Depending on the threshold of possible earthquake triggering, an increase of CFS by 50 KPa could be adequate to trigger subsequent earthquake events (HARDEBECK et al., 1998; see section 2.3.3).

5.6   Discussion of modelling results and conclusions

The results of the *sub-model A* study shows that the ABAQUS® *submodelling* technique is an adequate procedure for investigating the tectonic behaviour of a geological system at different geometric scales. In general, the modelled state of stress obtained from the *sub-model A* FEA is in agreement with the results of the global model study. Thus, the ABAQUS® *submodelling* technique enables the transfer of loading conditions for a geological system from a regional (i.e. global) model onto local model domains with a higher spatial and geometric resolution (i.e. sub-model(s)). Analysing these local model domains can provide details about the tectonic behaviour of single fault structures under the modelled conditions.

The modelling results suggest that the occurrence of strain partitioning along implemented faults is a function of both the geometric properties of each individual fault surface (i.e. orientation and surface area) and the structural setting of that individual fault segment. The results of *sub-model A* predict a segment of the Eastern Border Fault south of the city of Rastatt to be the most active fault segment within the central segment of the URG. This prediction is supported by relative surface displacements and the calculated dilation tendency and slip tendency values (Figure 5.7b; Figure 5.8). Unfortunately, no geological evidence exists that either proves or disproves this observation.

The observed disparity between the vertical displacement pattern obtained from the global model results and the subsidence pattern within the URG (Figure 4.2.1; Figure 5.6 a) is not observable in the *sub-model A* results. For the upper kilometres of the model the faulting mechanisms predicted for the individual fault surfaces are highly variable with inversion, strike-slip and extensional reactivation predicted. This is in agreement with focal mechanism solutions and reflection seismic sections for the study area (AHORNER et al. 1983; LARROQUE et al., 1987; DELOUIS et al., 1993; PLENEFISCH & BONJER, 1997; WIRSING et al., 2007).

The results of the dilation tendency and slip tendency analysis reveal that these parameters can only be used as relative measures of the likelihood of reactivation of particular fault surfaces. It is thus not possible to determine whether shear or tensile failure is the more likely mode of reactivation using these parameters alone. It is interesting to note that for the central segment of the URG the location of the majority of damaging earthquakes coincides with fault segments characterised by low values of both slip tendency and dilation tendency (Figure 5.15).

The DT and ST calculated represent the static loading conditions of the modelled fault system. The results of the risking parameter analysis can be used to determine whether fault segments are relatively prone to reactivation. However, this method is not capable to predict the character of reactivation (i.e. creep or stick slip) or the affected surface area. Since the analysis of these risking parameters describes the relative likelihood of failure of a fault surface within an imposed stress field no dynamic processes such as earthquake activity can be addressed using this approach.

In order to address this dynamic behaviour for a single event an earthquake model analysis was conducted for the 1933 Rastatt event which occurred on a fault characterised by low slip tendency (Figure 5.15). Benchmarking data shows that the pseudo-dynamic model correctly estimates the magnitude of the earthquake (Table 5.4). In addition, the model reveals a similar faulting mechanism at the same spatial position compared to the Rastatt event. Providing that the angle between the fault surface and the orientation of maximum principal stress component is known, the stress field can be derived from the modelled displacement vectors (Figure 5.12 b). The results of the pseudo-dynamic earthquake model also indicate that the modelled stress state is a valid approximation of the present-day stress field of the URG. This further implies that the loading procedure used to drive the sub-model is appropriate.

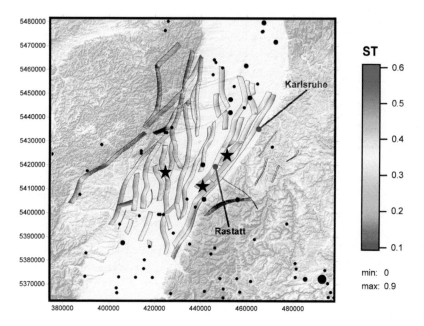

Figure 5.15: Instrumental earthquakes with magnitude M_L 2 – 5.4, documented between 1933 and 2002 (LEYDECKER, 2005A) compared with the modelled slip tendency results. Stars indicate earthquakes with magnitudes larger than 5 (Rastatt, 1933; Forchheim, 1948; Seltz, 1952).

PREDICTING THE IN-SITU STRESS STATE AND THE FAULT REACTIVATION POTENTIAL IN THE NORTHERN SEGMENT OF THE URG (SUB-MODEL B)

6.1   Introduction and structural setting

The northern segment of the URG is characterised by continued subsidence since the early Miocene (see section 1.3.2). The considerable thickness of Pleistocene and Quaternary sediments within the northern segment indicate a relative increase in accommodation space development during this time interval, presumably in response to increased tectonic activity. During the Pleistocene and Quaternary the proposed sinistral reactivation of the graben system and its bounding faults results in the northern and southern segments of the URG forming a releasing bends. Both the northern and southern URG and its bounding faults are oriented more favourably for present-day reactivation than the central segment (Figures 4.25 and 4.27). This leads to the suggestion that the imposed loads are more likely to be dissipated by movements along pre-existing fault structures. This interpretation is supported by the distribution of Quaternary sediments within the graben, which shows an increase in thickness with its distinct maximum in the Heidelberger Loch (Figure 1.4 B). The northern segment of the URG is characterised by a few scattered earthquakes only (BONJER et al., 1984), whereas damaging earthquakes have occurred less frequently in historical times as compared to the central segment (Figure 1.14).

The main focus of the second local scale modelling study is to investigate the reactivation potential of pre-existing faults in the northern segment of the URG under the present-day stress field. Furthermore, surface displacements and relative fault slip are investigated and modelled fault kinematics is compared with focal mechanism determinations. For the northern graben segment a detailed data base of reference data exists from the integrated study of palaeoseismology and geomorphology that was conducted by PETERS (2007). This data is used in conjunction with other data sets to evaluate the modelling result for the *sub-model B* volume.

6.2   Construction of sub-model B

The 3D Finite Element model (FEm) comprises the northern segment of the URG, two marginal basins, the Mainz Basin and Hanau-Seligenstädter Senke, and the adjacent shoulder units. Analogous to *sub-model A* (Chapter 5), *sub-model B* contains two lithospheric layers: the lower crust and the upper crust consisting of shoulder, graben and Cenozoic sedimentary infill (Figure 6.1). Due to the limited resolution of the model a uniform thickness of Cenozoic sediments of 3 km is assumed. The same technical procedure of discretisation was used as described in section 5.2.

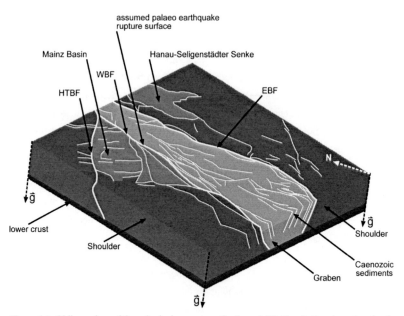

Figure 6.1: Oblique view of the spherical geometry of *sub-model B*. Gravitational acceleration is acting towards the centre of the Earth. The model dimensions are approximately 120 x 113 km, and the total thickness of the model is approximately 20 km. The model comprises two lithospheric layers, three different upper crustal units and a simplified fault model (white lines) of first and second order fault surfaces implemented as frictional contact surfaces (see section 3.1.2; WBF, Western Border Fault; EBF, Eastern Border Fault; HTBF, Hunsrück Taunus Border Fault). For references of the fault model used see Figure 1.9. In addition, the location of the postulated palaeo-earthquake at the WBF is given.

Analogous to the previously described finite element models, lateral depth dependent density variations are assigned to the lower and upper crustal units of *sub-model B*. These density variations are derived from Bouguer Anomaly data and are assigned as a field dependent variable to the nodes of the model (see Figure 4.2). First order tetrahedral elements are used; the specifications on the element types of the various units are given in Table 6.1.

Model Unit	Element Type	~Element Size [m]	~Quantity
Cenozoic sediments	C3D4	700	166,000
Graben & marginal basins	C3D4	700-2000	618,000
Shoulders	C3D4	700-2000	890,000
Lower crust	C3D4	2000	39,000

Table 6.1: First order tetrahedral elements are used for the discretisation of the various model units. A maximum resolution of 700 m is achieved within the graben and along the frictional contact surfaces implemented. Average element size and quantity are presented for each unit.

6.3   Modelling parameters and boundary conditions

Analogous to the analysis of *sub-model A*, the appropriate boundary nodes of *sub-model B* are constrained by displacements interpolated from the global model. Additionally, similar rheological properties are used for the various units of the sub-model. The deformation within the upper crustal units is also described using elastic material behaviour. The rheological properties of the units implemented are given in Table 6.2.

Model Unit	Rock Type	Density [kgm^{-3}]	Young's Modulus [GPa]	Poisson's Ratio	Reference
Cenozoic sediments	Sediments	2450-2550	30	0.275	TURCOTTE & SCHUBERT, 2002
Graben & marginal basins	fractured basement	2450-2750	45	0.275	TURCOTTE & SCHUBERT, 2002
Shoulder	Sandstone, Granite and Gneiss	2550-2750	55	0.275	TURCOTTE & SCHUBERT, 2002
Lower crust	Anorthosite	2750-2850	83	0.4	TURCOTTE & SCHUBERT, 2002

Table 6.2: Elastic rheological properties assigned to the 3D FEm units.

6.4 Loading procedure

The loading procedure adopted for *sub-model B* was similar to the loading procedure of *sub-model A* (see section 5.4).

6.4.1 Gravitational pre-stressing model

In this first step the model is only loaded by gravitational acceleration towards the Earth's centre and is basally and laterally constraint by a displacement field interpolated from the gravitational pre-stress step of the URG global model (Figure 6.2; see section 4.4.1).

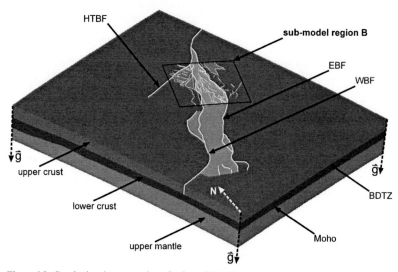

Figure 6.2: Gravitational pre-stressing of *sub-model B*. The model is loaded only by body forces induced by gravitational acceleration towards the Earth's centre. In order to achieve a model with relatively increasing horizontal stresses towards the Earth's surface, the model is basally and laterally constraint by nodal displacements interpolated from the pre-stressing step of the URG global model. WBF, Western Border Fault; EBF, Eastern Border Fault; HTBF, Hunsrück Taunus Border Fault.

6.4.2 Tectonic pre-stressing model

Similar to the modelling procedure for the *sub-model A* analysis, the stress tensor obtained after the gravitational pre-stressing step is used as initial stress conditions for the tectonic pre-stressing step. This gravitationally loaded model is then laterally deformed by transferring nodal displacements from the calibrated and benchmarked tectonic pre-stressing step of the global model onto the appropriate basal and lateral boundary nodes of the sub-model (Figure 6.3).

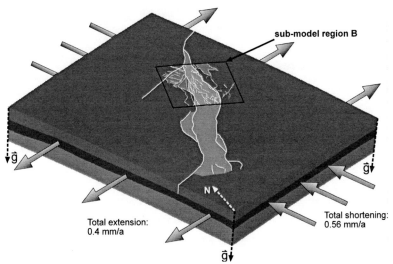

Figure 6.3: Oblique view of the tectonic loading of *sub-model B*. Arrows indicating the simplified lateral loading conditions applied perpendicular to the global model boundaries. Nodal displacements obtained during the global model analysis are transferred on the appropriate basal and lateral nodes of the sub-model region.

6.4.3 Modelling the short-term deformation pattern of the central URG

In order to address the short-term deformation pattern of the northern URG area, the boundary conditions obtained from the global model for a time increment of 10 ka are transferred on the sub-model boundaries. Analogous to the global model, the vertical surface displacements obtained are compared to uplift data derived from geological and geomorphological data. In addition to the surface uplift predictions, modelled fault slip rates are compared to available fault slip rate data.

6.4.4 Modelling the static stress transfer (ΔCFS) induced by a possible palaeo-earthquake in the northern segment of the URG

In order to analyse the static stress transfer induced by a possible palaeo-earthquake located at the Western Border Fault, the modelled stress tensor of the tectonically loaded *sub-model B* is used. Analogous to the modelling approach used for *sub-model*

A, at the time increment prior to the earthquake event, no relative fault movement is assumed for the entire model. At the time of the earthquake event, only relative slip along the rupture plane within the brittle upper crust is considered. The concept of ΔCFS is used to investigate whether fault planes in the vicinity of the earthquake event are transferred closer to failure (see section 2.2.3). The results of the ΔCFS analysis are also compared to the distribution of slip tendency along the modelled fault surfaces.

6.5 Modelling results

The purpose of *sub-model B* is to obtain an approximation of the present-day state of stress affecting the northern segment of the URG. The influence of the dense network of second and third order fault surfaces on the state of stress, the kinematic behaviour and the displacement pattern is investigated. In this section the modelling results for the northern segment of the URG are presented.

6.5.1 Prediction of the state of stress in the northern segment of the URG

The state of stress predicted by *sub-model B* is similar to the prediction of the global model (compare with section 4.5.2). In part, this is because, with the exception of the upper two kilometres of sedimentary fill, identical material properties are used. Furthermore, the complete loading procedure of the global model is transferred on the sub-model boundaries using the ABAQUS™ *submodelling* technique. For comparison, the modelling results are presented for a uniform reference depth of 3000 m below sea level.

A transtensional state of stress is predicted for the northern segment of the URG (Figure 6.4a). Variations in the transtensional state of stress are induced by spatially varying material properties, crustal thickness, surface topographies and differential loading of the faults. The sub-model predicts a similar RSR magnitude distribution to the global model (Figure 4.15). In contrast to the results of *sub-model A*, a general increase in RSR values is not predicted for the graben area by the *sub-model B* results. RSR magnitudes are elevated for the NW part of the model for both the shoulder and graben regions (RSR values of 0.9 to 1.2). In the vicinity of HTBF, the effect of ΔGPE on the predicted RSR values becomes obvious. At its northern margin, the Mainz Basin is affected by additional horizontal loading due to density and topography variations relative to the adjacent Rhenish Massif. Higher RSR magnitudes of the graben infill compared to the shoulder regions are only predicted for the southern part of the model. The reduced RSR magnitudes within the graben segment located to the west of the city of Heidelberg can be attributed to the orientation of the graben with respect to the imposed stress field. As predicted by the global model, in this area the horizontal stress components (S_H and S_h; Figure 4.13 and Figure 4.14) are reduced due to the dynamically imposed releasing bend setting (Figure 4. 27).

In order to qualify the relative likelihood of brittle failure predicted by *sub-model B* relative magnitudes of the second invariant of the stress tensor (i.e. von Mises stress, σ_M) are analysed. The sub-model results show a similar σ_M magnitude distribution to the global model (Figure 4.16). A relative increase in σ_M magnitude is predicted for the shoulder regions, whereas relatively decreased σ_M magnitudes are predicted for the graben region (Figure 6.4b). For the graben and sub-basin regions a lower elasticity is defined resulting in a more isotropic state of stress. In addition, reduced σ_M magnitudes are predicted along the faults considered. The distribution of predicted σ_M magnitudes illustrates the influence of the pre-existing Variscan structure of the crust on the state of stress of the northern part of the URG.

Figure 6.4a (next page): Regime stress ratio (RSR; see section 2.1.2) at 3000 m depth. In general, the model predicts a transtensional state of stress for the northern segment of the URG. First order fault structures, also implemented in the URG global model, are indicated by bold white lines. Thin white lines indicate second and third order fault structures only defined in the sub-model volume. Figure 6.4b: Von Mises stress magnitudes (σ_M; see section 2.1.2.3) at 3000 m depth. σ_M shows a strong dependency on the surface topography and the structural composition of the model volume. Within the graben σ_M magnitudes are relatively decreased compared to the shoulder regions with local variations occurring. Note the locally induced stress concentrators at the intersection of the first order fault structures implemented with the lateral model boundaries. Using a similar model geometry with the ABAQUS ™ *submodelling* technique, these stress concentrators are unavoidable. Fortunately, the stress concentrators are only significant locally.

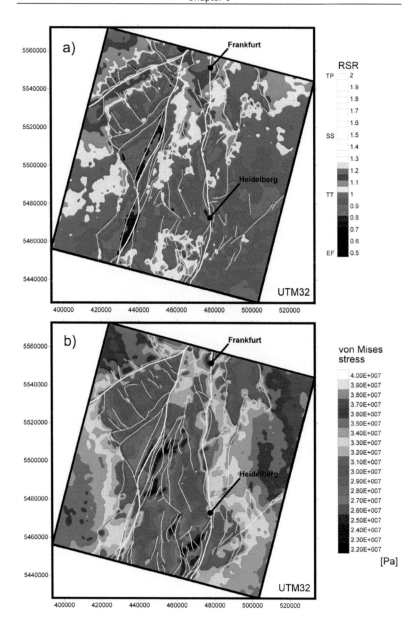

6.5.2 Prediction of relative surface displacement and fault slip magnitudes (sub-model B)

In order to analyse the induced deformation along upper crustal faults within the central segment of the URG, the displacements from the global model for a time span of 10 ka are applied as nodal constraints to the appropriate boundaries of *sub-model B*. The distribution of the surface displacement predicted is similar to the results of the global model (Figure 4.20). However, due to the higher fragmentation of the brittle upper crust resulting from the implementation of second and third order faults the sub-model predicts lower magnitudes of relative vertical displacement (Figure 6.5).

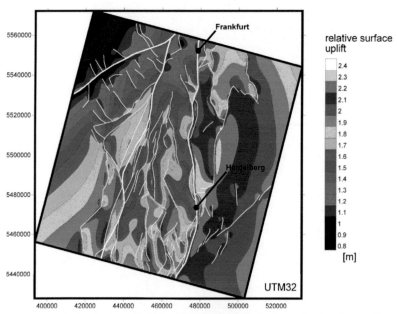

Figure 6.5: Calculated values of relative surface uplift induced. The predicted surface uplift is highest for the eastern shoulder region with a distinct locus of maximum relative surface uplift located in the region to the NE of the city of Heidelberg. This region corresponds to the position of the Odenwald basement high. Within the graben a complex vertical surface displacement pattern is predicted. Note the relative subsidence predicted for the region of the Rhenish Massif in the NW of the model.

The modelled displacement pattern suggests that increased relative uplift is induced along the eastern shoulder, whereas a distinct maximum of relative surface uplift is predicted for a region NE of the city of Heidelberg coincident with the location of the Odenwald basement high (Figure 6.5). Within the graben, the model predicts a complex displacement pattern with varying uplift and subsidence of individual fault blocks (Figure 6.5). Analogous to *sub-model A* results, the *sub-model B* results indicate varying fault kinematics under an overall transtensional state of stress. In contrast to the *sub-model A* results, inversion of extensional fault structures is less likely in the northern segment of the URG. In comparison to the global model results, *sub-model B* predicts a more realistic vertical displacement pattern for the intra-graben region with varying differential displacements along the implemented second and third order intra-graben faults. Similarly to *sub-model A*, the results of *sub-model B* indicate that the prediction of relative surface displacements depends primarily on the geometric complexity of the model.

The model shows a complex pattern of differential fault-block movements as suggested by the interpretation of precise levellings (PRINZ & SCHWARZ, 1970; SCHWARZ, 1974; DEMOULIN et al., 1995). The position of highest simulated differential displacement of shoulder and graben units is predicted in the vicinity of the city of Heidelberg and coincides with the locus of highest Quaternary subsidence within the URG, the Heidelberger Loch. Analogous to the global model results, the sub-model fails to predict the elevated uplift of the Rhenish Massif documented from measurements of river incision (MEYER & STETS, 1998; VAN BALEN et al., 2000; see discussion in section 4.5.4).

Figure 6.6 compares the modelled surface displacement pattern with the results of the geomorphological observations of PETERS & VAN BALEN (2007A). In general, the modelled uplift and subsidence pattern is in agreement with the field data. Furthermore, the location of highest activity along the Eastern Border Fault (i.e. uplift of Odenwald basement high, subsidence of Heidelberger Loch) is simulated successfully. For the Mainz Basin, geomorphological studies reveal that the largest recent uplift occurred on the southern border of the Niersteiner Horst (PETERS & VAN BALEN, 2007B).

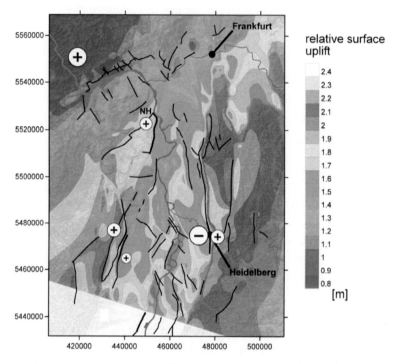

Figure 6.6: Modelled relative surface uplift compared to relative vertical movements' of Late Pleistocene age inferred from the palaeogeographic evolution and geomorphic indices calculations after PETERS & VAN BALEN (2007A). Fault segments with documented recent activity (~ Late Pleistocene to Early Holocene) are indicated with solid black lines, + relative uplift, - relative subsidence. NH, Niersteiner Horst.

The modelling study also predicts higher surface uplift for the fault blocks located to the south of the Niersteiner Horst. Both studies indicate that the Niersteiner Horst does not behave as a horst under the contemporary kinematic conditions. At the Western Border Fault, in the vicinity of the postulated palaeo-earthquake, extensional faulting is predicted by the model, which is documented from field observations (PETERS et al., 2005; Figure 6.1). The dominant uplift of the Rhenish Massif, documented from the elevation of river terraces, is not predicted by the sub-model (PETERS & VAN BALEN, 2007B) since the Eifel Plume responsible for the uplift has not been implemented (see discussion in section 4.5.4).

The modelled fault slip rates are now compared to slip rates determined for the URG border faults on the base of geological and geomorphological criteria (Table 6.3). The comparison shows good agreement between the predictions and observed data.

Location (No. in Fig. 4.23)		Period [ka]	Method	Reference	Fault slip rate [mm/a]	Fault slip rate of model after 10 ka [mm/a]
Northern URG	Western Border Fault near Osthofen (1)	800	Terrace mapping	PETERS & VAN BALEN, 2007A	0.06 – 0.13	0.02 – 0.05
	Western Border Fault near Osthofen (1)	8-19	Paleoseis-mological trenching	PETERS et al., 2005	0.04	0.03
	Eastern Border Fault at Heidelberger Loch (2)	1800	Sediment thickness	BARTZ, 1974	0.2	0.1

Table 6.3: Fault slip rates of the URG border faults obtained from geomorphological and geological investigations compared to modelling results of *sub-model B*. For the location of the fault segments see Figure 4.23.

In order to visualise the kinematic behaviour of the implemented fault surfaces, the magnitude of the gradient of the relative surface displacement is calculated (Equation 4.4; Figure 6.7a). The distribution of increased vertical displacements along the graben bounding faults shows several maxima. A distinct maximum of relative vertical displacement is predicted for a segment of the Eastern Border Fault (EBF) in the vicinity of the city of Heidelberg. This segment is characterised by a change in strike from the overall NE/SW trend to a N/S trend. For the EBF system in general, the model predicts relatively increased values of the relative surface displacement gradient, which is in agreement with the global model results (Figure 4.23). Similar to the EBF system, the relative displacement gradient of the Western Border Fault (WBF) system shows a dependency on the orientation of the particular fault segment. Within the graben, the model also predicts that NE/SW oriented fault structures are less likely to accommodate relative vertical displacements. As observed in both the global model and *sub-model A* results, the distribution of increased horizontal displacements is more uniform than the vertical component. The relative horizontal displacement gradient has a fewer dependants on the orientation of each particular fault segment (Figure 6.7b). Similar to the predictions of the relative vertical displacement gradient, the gradient of relative horizontal displacement is relatively increased for the EBF to the North of the city of Heidelberg. Furthermore, the gradient of relative horizontal displacement is relatively increased for the northernmost segment of the WBF where this fault structure separates the Mainz Basin from the northern URG (Figure 6.7b).

In order to visualise the total relative fault slip predicted by *sub-model B*, the vertical and horizontal displacement gradients have been normalised and combined in Figure 6.8. The modelling results indicate that for the northern segment of the URG the relative fault slip predicted is highest for a segment of the EBF in the vicinity of the city of Heidelberg. This location coincides with the locus of largest Pleistocene to Quaternary subsidence within the URG (i.e. Heidelberger Loch; Figure 1.4).

Figure 6.7a (next page): Calculated values of relative surface displacement gradient. Fault segments characterised by an increased dip-slip component (i.e. extensional faulting) are indicated by an elevated relative surface vertical displacement gradient. Figure 6.7b: Calculated values of the N/S component of the relative surface displacement gradient. Fault segments characterised by an increased strike-slip component are indicated by an elevated relative horizontal surface displacement gradient.

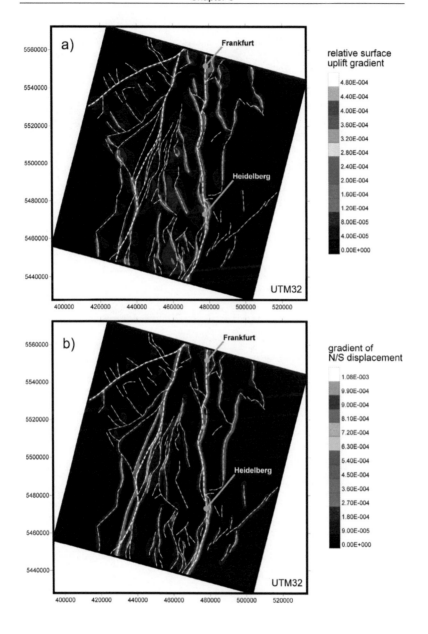

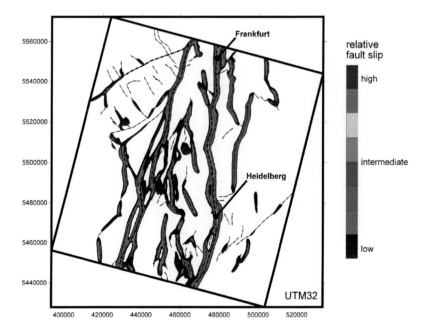

Figure 6.8: Relative fault slip along the implemented fault model in the northern segment of the URG derived from surface displacement gradients.

6.5.3 Dilation and Slip tendency analysis for sub-model B

In this section, the dilation tendency (DT) and slip tendency (ST) results for upper crustal faults of the *sub-model B* domain are presented and discussed. The theory of the DT and ST parameters is described in section 2.3 and a short introduction of their principles and assumptions given in section 5.5.3.

Figure 6.9a shows the calculated values of the dilation tendency for the *sub-model B* domain. DT values in the range of 0 ($\sigma_n = \sigma_1$) to 1 ($\sigma_n = \sigma_3$) are predicted for the fault model of the northern segment of the URG. Relatively increased values of DT are predicted for the vertical, NNE/SSE and NW/SE trending fault structures in the eastern shoulder region and in the Mainz Basin due to their orientation relative to the postulated transtensional state of stress. Furthermore, extensional faults throughout the model volume, including the URG and its marginal basins (i.e. Mainz Basin and Hanau Seligenstädter Senke), with NNW/SSE to NNE/SSW oriented strikes show increased DT values. High DT values are predicted for a segment of the EBF to the

North of the city of Heidelberg. In contrast, NE/SW striking Variscan terrane boundaries and associated fault structures generally show relatively decreased values of DT (e.g. HTBF).

Figure 6.9b shows the calculated values of slip tendency for the *sub-model B* domain assuming a static coefficient of sliding friction of μ_{static} = 0.4. In general, ST is increased for NNW/SSE to NNE/SSW oriented extensional faults throughout the model volume. Analogous to the dilation tendency, slip tendency is relatively decreased for NE/SW striking Variscan terrane boundaries and associated fault structures. Similar to the DT results, the same segment of the EBF to the North of the city of Heidelberg shows increased magnitudes of ST. In contrast to the DT results, vertical faults generally show relatively decreased ST magnitudes due to their unfavourable orientation relative to the state of stress.

Figure 6.10 relates the DT and ST values to the distribution of fault strikes for the *sub-model B* domain. In general, the comparison of the results of both sub-models reveal that within the northern and central segments of the URG, relatively increased values of both DT and ST are predicted for N/S striking extensional structures (Figure 6.10). Analogous to the analysis of the *sub-model A* results the limitation of the method prevents determination whether tensile or shear is failure is the most likely mode of reactivation of upper crustal faults within the northern URG area. The results of the combined DT and ST analysis indicate that the majority of extensional fault structures in the northern URG, established during the Eocene/Oligocene extensional phase, are favourably oriented, relative to the present-day stress field, for both tensile and shear reactivation (i.e. transtensionally; 0.5 < RSR < 1.5).

Figure 6.9a (next page): Calculated magnitude of dilation tendency (DT) for upper crustal faults within the northern segment of the URG within 4 to 8 km depth. Figure 6.9b: Calculated magnitude of slip tendency (ST) for upper crustal faults within the northern segment of the URG over a depth range of 4 to 8 km, assuming a sliding friction coefficient of $\mu_{static} = 0.4$.

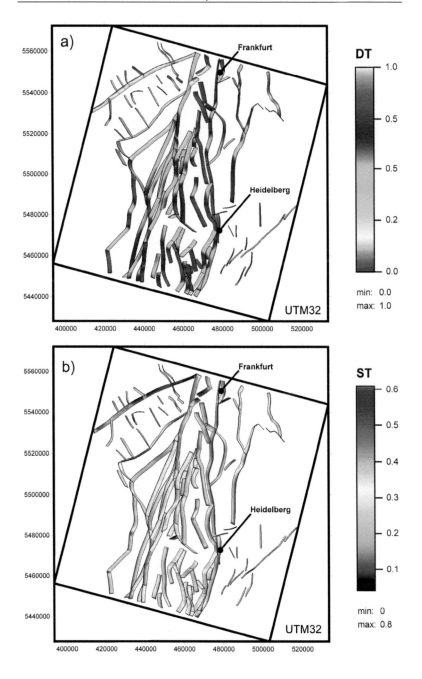

Figure 6.10 a1) (next page): Rose diagram of fault strike orientations for upper crustal faults of the northern segment of the URG compared to the distribution of predicted DT values. Increased DT values (≥ 0.5) are predicted for fault segments with strikes ranging from N340° to N10°. a2) Histogram indicating the distribution of DT values for the individual surface elements for *sub-model B*. The histogram shows a uniform distribution of DT values in the range between 0.45 and 1. The mean value of DT is 0.66. When compared to the *sub-model A* results, DT is slightly elevated in the northern segment of the URG. b1) Rose diagram of fault strike orientations for upper crustal faults compared to the distribution of predicted ST values. Relatively increased ST values (≥ 0.3) are predicted for faults with strikes ranging from N340° to N20°. b2) Histogram indicating the distribution of ST values for the individual surface elements. The histogram shows a distinct maximum at 0.36 for the ST distribution. The mean value of ST is 0.32. Compared to the *sub-model A* results, ST is slightly reduced in the northern segment of the URG.

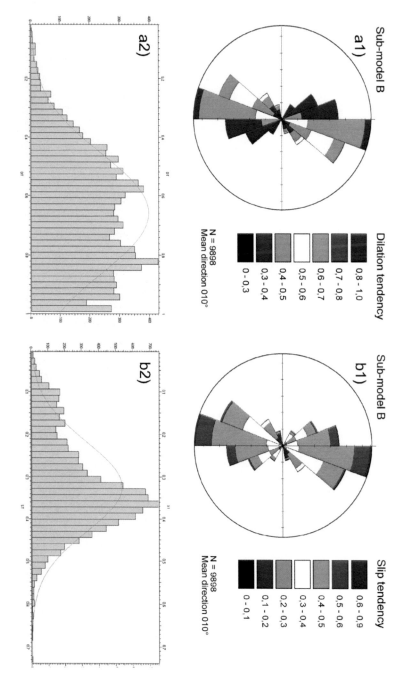

6.5.4 Static stress transfer (ΔCFS) induced by a possible palaeo-earthquake in the northern URG

In this section the static stress transfer induced by a possible palaeo-earthquake along the northern segment of the Western Border is investigated using a similar modelling approach to the one used for the Rastatt 1933 earthquake (see section 5.5.4). PETERS et al. (2005) identified deformation of near surface sediments at a location along the WBF in trenches that were excavated for palaeoseismological studies. The authors concluded that the observed deformation structures may be the result of a palaeo-earthquake. Since unambiguous proof of a seismic event is only possible under rare circumstances, aseismic creep is also likely to have caused the observed structures. For the modelling study, it is assumed that the observed maximum relative surface displacement of 0.5 m is the result of one seismic event. Following this assumption, PETERS et al. (2005) derived a possible moment magnitude (M_W) of 6.0 to 6.5. In historical times no earthquake of this magnitude has been observed in the northern URG. An exception exists for the southern URG in the 1356 Basel earthquake, which has an estimated magnitude of 6.0 to 6.6 (e.g. AHORNER & ROSENHAUER, 1978; MAYER-ROSA & CADIOT, 1979; MAYER-ROSA & BAER, 1992). Further evidence for palaeo-earthquakes in this magnitude range comes from palaeoseismological investigations in the area of Strasbourg in the central segment of the URG. CUSHING et al. (2000) identified faulted Quaternary deposits in two palaeoseismological trenches and postulated several palaeo-earthquake events in the magnitude range of 6.0 to 6.5.

For the study presented herein, the mean moment magnitude of 6.2 for the possible palaeo-earthquake located at the WBF is used to calculate the rupture surface area, which is then used to define the rupture surface within the FE model. Based on the empirical approach of WELLS & COPPERSMITH (1994), the geometric properties of the modelled rupture plane are given in Table 6.4.

hypocentral depth	height	length	area
6 km	12 km	12 km	144 km^2

Table 6.4: Geometric properties of the modelled rupture plane. The hypocentre is assumed to be in the centre of the rupture plane. On average, the rupture plane dips with 65° and has a dip direction of 100°. In contrast to the Rastatt 1933 earthquake the possible palaeo-earthquake is assumed to have caused a surface rupture, which was detected by palaeoseismological investigation.

Similarly to *sub-model A*, the stress tensor obtained from the tectonically loaded *sub-model B* is used to analyse the static stress transfer induced by the palaeo-earthquake. Within the earthquake model, slip is allowed to occur only on the predefined rupture plane. Using the concept of ΔCFS, it is investigated whether fault planes in the vicinity of the earthquake event are transferred closer to failure (see section 2.2.3). Additionally, the results of the ΔCFS analysis are compared to ΔST results (change in slip tendency due to the tectonic event) and the distribution of ST modelled along the faults. The principle assumptions for the calculation of ΔCFS are given in section 5.5.4.

In order to assess the model quality, the modelled displacement of the implemented fault surface is used to calculate the moment magnitude (M_W) of the modelled palaeo-earthquake. The modelling results provide information on the average fault slip (D), whereas the rupture area (A) and the shear modulus (G) are defined as boundary conditions for the model analysis. Using these parameters, the seismic moment (M_0) can be calculated following the approach of KANAMORI & ANDERSON (1975; Equation 5.1). The seismic moment (M_0) is then used to determine the moment magnitude (M_W; Equation 5.2). Table 6.5 compares the moment magnitude derived from the modelled fault slip to the moment magnitude derived from the observed maximum relative surface displacement given by PETERS et al. (2005).

M_W derived from model input data (offset)	M_W derived from model results (D)
6.2	6.17

Table 6.5: Moment magnitude derived from the modelled fault slip compared to the moment magnitude derived from the available input data.

Figure 6.11 illustrates the construction and benchmarking procedure of the possible palaeo-earthquake model. The maximum offset along the rupture surface given by PETERS et al. (2005) is used to derive the area of the rupture plane and the moment magnitude for the possible earthquake event.

Input data

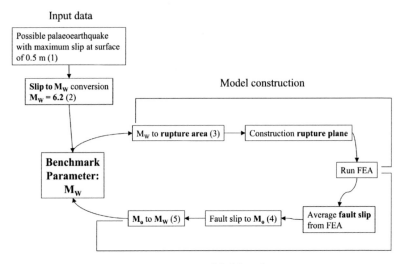

Model construction

Model results

Figure 6.11: Diagram illustrating the construction and benchmarking procedure of the possible palaeo-earthquake model. References for the conversion steps are given by numbers: (1) PETERS et al., 2005; (2) WELLS & COPPERSMITH, 1994; (3) WELLS & COPPERSMITH, 1994; (4) KANAMORI & ANDERSON, 1975; (5) STEIN & WYSESSION, 2003.

Similar to the Rastatt 1933 earthquake model, the displacement along the fault surface is used to calculate a pseudo fault plane solution. Figure 6.12 illustrates the results of the NDA for the simulated earthquake event. Since both the orientation of the fault plane and the orientation of the stress field are known from the modelling results, a realistic angle ($\theta = 45°$) between the reactivated rupture plane and the orientation of the maximum principal stress component is used to calculate the pseudo fault plane solution. The results of the NDA predict an extensional faulting mechanism with a minor strike-slip component. This is compatible with the sediment deformation observed in the palaeoseismological trenches (PETERS et al., 2005).

Sub-model B

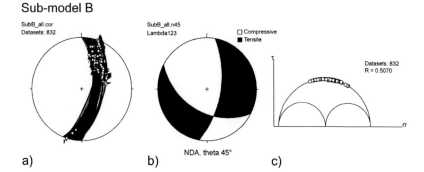

a)   b)   c)

Figure 6.12: a) Three dimensional distribution of the slip vector on the rupture plane, fault lineation data plotted on lower hemisphere, equal area projection. b) Pseudo fault plane solution derived from the modelled slip vectors for the possible palaeo-earthquake event using NDA ($\theta = 45°$). c) Calculated three dimensional Mohr circle for the fault slip data. The results of the NDA reveal a predominantly extensional faulting mechanism with a minor strike-slip component for the modelled earthquake.

The benchmarked *sub-model B* is now used to investigate the effects of the possible palaeo-earthquake on the stress field and possible triggering of sub-sequent earthquake events on adjacent fault surfaces. Figure 6.13 illustrates the induced surface displacement and change in maximum shear stress. The induced surface displacement shows a distinct asymmetry with, similar to *sub-model A* results, greater magnitudes occurring in the structural hanging wall (graben block). Since the palaeo-earthquake is in the order of one magnitude larger than the Rastatt 1933 earthquake, the surface area affected is significantly larger (Figure 6.13a). The induced surface displacement is a function of the faulting mechanism and the structural composition of the affected volume. The pattern of the induced change in maximum shear stress magnitudes shows an asymmetric distribution with a distinct maximum east of the rupture surface, in the graben centre. The asymmetric distribution of the change of this scalar value indicates a combination of strike-slip and extensional mechanisms for the palaeo-earthquake (compare Figure 6.12).

Figure 6.13 (next page) a: Surface displacement pattern induced by the simulated possible M 6.2 palaeo-earthquake. Note the asymmetry of the surface displacement pattern. Figure 6.13 b: Change in the maximum shear stress magnitude at a depth of 4 km induced by the palaeo-earthquake. The transparent yellow surface indicates the modelled rupture plane. Contour lines represent a change in maximum shear stress magnitudes of 100,000 Pa. Note the distinct locus of increased maximum shear stress to the SE of the simulated earthquake event.

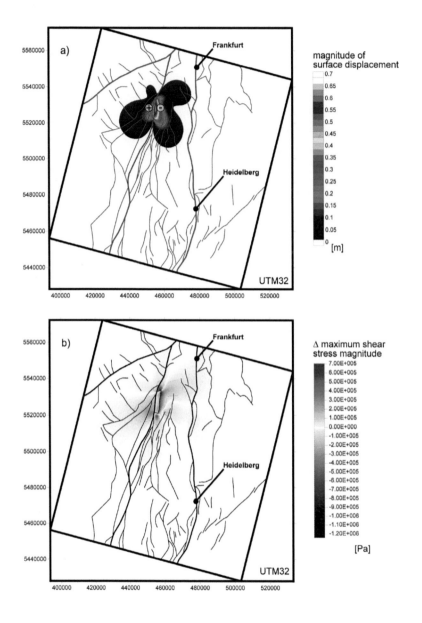

Analogous to *sub-model A*, the change in the maximum shear stress magnitude (τ_{max}) is used to illustrate the volumetric change of the stress field induced by the palaeo-earthquake (Figure 6.13 b). ΔCFS is only calculated on fault surfaces used to define the *sub-model B* geometry and the absolute change in shear stress is considered (Figure 6.14 b). In addition, the change in slip tendency (ΔST) is calculated for the surfaces of the fault model (Figure 6.14 a). The ΔST values show the change in the relative likelihood of fault slip. The results depend on the orientation of the fault surface relative to the stress field. ST is reduced significantly in the vicinity of the rupture surface. Adjacent to the rupture surface ST is increased on the same fault surface. The ΔST analysis reveals a greater influence of the possible palaeo-earthquake on the state of stress along adjacent fault surfaces than the results of the Rastatt 1933 earthquake model (compare Figure 5.14a and Figure 6.14a). ST is increased for a fault set situated in the SE of the simulated earthquake event (Figure 6.14a). In contrast to the relative approach of ΔST, the ΔCFS values show the absolute change in Coulomb Failure Stress (CFS) magnitudes for adjacent fault surfaces induced by the slip along the rupture surface. CFS is reduced by up to ~2.7 MPa in the vicinity of the rupture surface. Additionally, CFS is reduced significantly for marginal fault structures in the western shoulder region to the west of the modelled event (Figure 6.14 b). Adjacent to the rupture surface CFS is increased on the same fault surface by up to ~0.6 MPa. Furthermore, CFS is increased for a set of NNE/SSW striking faults within the URG situated in the east of the modelled event and for the northern boundary fault of the Niersteiner Horst situated NW of the modelled event (Figure 6.14 b). For both regions CFS is increased by up to 0.3 MPa. An increase of CFS by 0.3 MPa could be sufficient to trigger subsequent earthquake events (HARDEBECK et al., 1998; see section 2.3.3).

Figure 6.14 (next page): Change in planar parameters on the implemented fault model induced by the simulated palaeo-earthquake shown between 4 to 8 km depth. a: Change in slip tendency (ΔST) induced by the simulated palaeo-earthquake. b: Change in Coulomb Failure Stress (ΔCFS) induced by the simulated palaeo-earthquake.

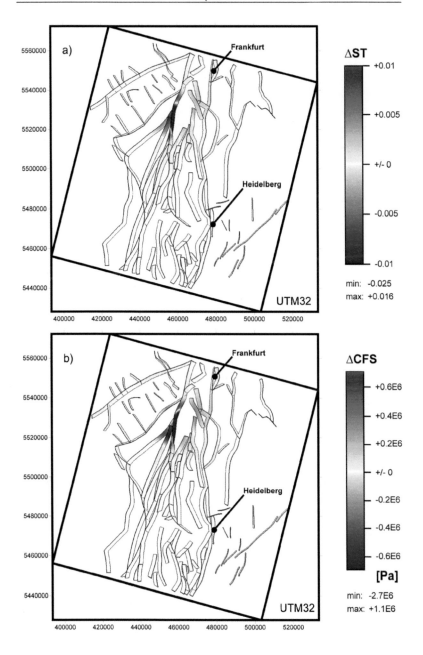

6.6 Discussion of modelling results and conclusions

Sub-model B results suggest that strain partitioning along the fault surfaces considered depends on their geometric properties and structural setting. The most active segment within the northern URG, similarly to the central URG, is predicted to be located at the Eastern Border Fault (EBF). The model suggests the highest probable tectonic activity for a segment of the EBF in the vicinity of the city of Heidelberg. This is suggested both by the slip and dilation tendency analyses and the predicted differential uplift in this area (Figure 6.6; Figure 6.9). According to geological and geomorphological data this segment is identified to feature large Quaternary displacements (BARTZ, 1974; PETERS & VAN BALEN, 2007A, HAIMBERGER et al., 2005).

The predicted surface displacement pattern for both the shoulder and graben regions is in accordance with observations based on precise levellings. These indicate a complex displacement pattern of individual fault blocks within the northern URG, whereas in general the shoulder regions are uplifted relative to the graben (KUNTZ et al., 1970; PRINZ & SCHWARZ, 1970; SCHWARZ, 1974; DEMOULIN et al., 1995). For the upper kilometres of the model the predicted faulting mechanisms for the individual fault surfaces are highly variable since the model predicts inversion, strike-slip and extensional reactivation to occur. This is in agreement with focal mechanism solutions for the study area (AHORNER et al. 1983; LARROQUE et al., 1987; DELOUIS et al., 1993; PLENEFISCH & BONJER, 1997).

Figure 6.15 compares the historical and instrumental seismicity with the slip tendency values calculated for the fault model of the northern URG. In this area damaging earthquakes have occurred less frequently compared to the central and southern graben segments. Similar to the results of sub-model A the ST analysis provides no constraints on the seismicity of an individual fault surface. For example, seismic activity is documented for the HTBF for which the modelling results suggest significantly reduced ST values (AHORNER & MURAWSKI, 1975). Furthermore, the modelling results show that the strongest documented earthquake within the northern URG in the vicinity of the city of Worms might be associated with a fault structure characterised also by reduced ST values (Figure 6.15).

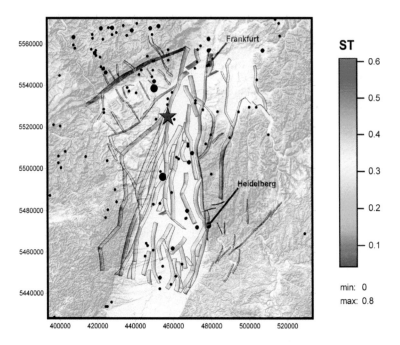

Figure 6.15: Historical and instrumental earthquakes with magnitude M_L 2 – 5.1, documented between 858 and 2002 (LEYDECKER, 2005A) compared with the modelled slip tendency results. The star indicates the location of the postulated palaeo-earthquake event .

In contrast to the Rastatt 1933 earthquake model, the tectonic event investigated for the *sub-model B* domain is located on a fault segment with relatively increased slip tendency values. In order to investigate the dynamic behaviour of the northern URG for a single tectonic event, a possible palaeo-earthquake located at the WBF is simulated. The scaling relation of the estimated rupture area and the predicted fault slip is considered by using ABAQUS® for the case of a fault surface that is optimally oriented for reactivation in the modelled stress field. The model correctly estimates the magnitude of the simulated earthquake (Table 6.5). In addition, the model reveals a similar faulting mechanism at the same spatial position compared to the structural analysis of fault data at the location of the palaeoseismological investigation (PETERS et al., 2005). Analogous to the results of the Rastatt 1933 earthquake model, the stress field can be derived from the modelled displacement vectors given that the angle between the orientation of the maximum principal stress component and the rupture plane is known (Figure 6.12b). The results of the pseudo-dynamic earthquake model indicate that the modelled stress state is a valid approximation of the present-day stress field of the northern URG. This further implies that the loading procedure used to drive *sub-model B* is appropriate.

INTEGRATED DISCUSSION AND CONCLUSION OF THE MODELLING RESULTS

7.1 Objectives of this thesis

The objective of the research presented in this thesis was to simulate the contemporary 3D crustal state of stress for the Upper Rhine Graben and the adjacent shoulder regions. Based on this simulation the implications of the imposed kinematics and dynamics on the dense network of pre-existing fault structures within the URG area were investigated. The state of stress and the kinematic behaviour in the area is, amongst others, the result of the interaction of the complex crustal structure, the imposed plate boundary forces (i.e. Atlantic ridge push and Alpine collision related compressional forces; e.g. GRÜNTHAL & STROHMEYER, 1992; GÖLKE, 1996) and local thickness and density variations resulting in gravitational potential energy differences. In order to honour the influence of both a complex distribution of internal properties and external (i.e. imposed) loads on the state of stress and the kinematic behaviour affecting the URG geological system, the 3D finite element modelling method was chosen as the appropriate simulation technique. Knowledge of the 3D state of stress is essential both for the understanding of geodynamic processes as well as for the evaluation of possible reactivation of individual fault surfaces. Furthermore, knowledge of the 3D state of stress is vital to optimise development and production processes in anthropologically important situations such as hydrocarbon and geothermal systems. The two main objectives of the study presented herein are:

- To develop a series of multi-scale finite element models to approximate the full crustal stress tensor and the kinematic behaviour for the URG area. These models are applied to the URG scale (i.e. global model) and to scales comprising the central graben segment (*sub-model A*) and the northern graben segment (*sub-model B*) respectively.

- To assess the reactivation potential of pre-existing first and second order faults within the URG area under the approximated 3D state of stress.

7.2 The finite element methodology

In order to achieve the study objectives, a multi-scale modelling approach based on the combination of the ABAQUS™ *submodelling* technique (see section 3.2.1), a spherical loading procedure (see section 3.3.2) and a pre-stressing approach (see section 3.3.1) was developed. The purpose of this approach is to enable the approximation of the in-situ state of stress for an arbitrarily chosen geological volume integrating both large scale tectonic processes and local scale geometries and heterogeneities. The ABAQUS™ *submodelling* technique is a tool that enables the influence of the imposed plate scale loading conditions at a regional scale to be studied at the local scale using a series of models such as the URG global model and the two sub-models presented in chapters 5&6. The advantage of this approach is that it can be applied over various scales on a local model domain, for example the northern URG, where local boundary conditions are difficult or impossible to distinguish. Furthermore, the ABAQUS™ *submodelling* technique can be used to reduce the effect of pre-defined displacement boundary conditions on the modelling results within the area of interest since the location of these imposed boundary conditions can be separated spatially from the interpreted model domain.

The most important factor in simulating the contemporary in-situ state of stress of a geological volume is the application of gravitational acceleration. Similar to the conclusions of previous modelling studies the modelling study presented herein shows that using only elastic material properties in a gravitational accelerated model is insufficient to simulate a state of in-situ stress commonly observed within the Earth's crust (e.g. BUCHMANN & CONNOLLY, 2007; PETERS, 2007; ECKERT, 2007; HEIDBACH et al., submitted). This model set-up fails to generate sufficiently large horizontal stress magnitudes to result in the commonly observed extensional to compressional tectonic regimes (Figure 3.6). A spherical loading procedure and a pre-stressing approach is presented in this thesis in order to avoid unrealistic elastic compaction of the model domain and unrealistic predictions of an in-situ state of stress, which is contrary to observation, when using elastic material properties. Furthermore, this approach enables to address the so-called near surface horizontal stress paradox (ENGELDER, 1993; see section 3.3.2), which describes the commonly observed relative increase of the horizontal stress components towards the Earth's surface (Figure 3.6). This observation cannot be simulated directly using models with perpendicular, Cartesian coordinate systems. In contrast to the modelling approach suggested by ECKERT (2007), HEIDBACH et al. (submitted), the loading procedure developed during this thesis yields initial in-situ stress conditions that equilibrate the gravitationally induced loading, account for the relative increase of the horizontal stress components towards the Earth's surface and that are consistent with the observations from in-situ stress determinations without introducing additional model domains. The multi-scale modelling procedure comprises the following steps (Figure 7.1):

- A calibrated generic spherical model at the spatial position of the subsequent models is used to derive the necessary elastic compaction to account for the so-called near surface horizontal stress paradox (see section 3.3.2).

- The results of the generic spherical model are used to interpolate displacement boundary conditions for the gravitational loading step of the URG global model (see section 4.4.1).

- The stress state obtained from the gravitational loading step at each element integration point is included in the tectonic loading step of the global model during which lateral boundary conditions inferred from the World Stress Map data base are used to simulate the far-field loading of the URG (see section 4.4.2).

- At a local (i.e. sub-model) scale the results of the individual loading steps of the global model are used to interpolate displacement boundary conditions for the appropriate loading step of the individual sub-models (see section 5.4 and 6.4).

Figure 7.1 (next page): Concept of the ABAQUS™ *submodelling* procedure. In the multi-scale modelling approach presented herein, an initial spherical model is used. This simplified two layered model represents a part of the Lithosphere of the Earth. This generic spherical model is gravitationally accelerated towards the Earth's centre (model 1). The basal and lateral displacements obtained are then applied as boundary conditions on the global model (model 2). After this gravitational pre-stressing step, simplified lateral displacement boundary conditions are applied on model 2 (i.e. tectonic pre-stressing). The obtained basal and lateral displacements from the gravitational pre-stressing step of the global model are then applied as boundary conditions on the sub-model (model 3). After this gravitational pre-stressing step of the sub-model, the lateral displacements obtained from the tectonic pre-stressing step of the global model are used as boundary conditions for the tectonic pre-stressing step of the sub-model (model 3).

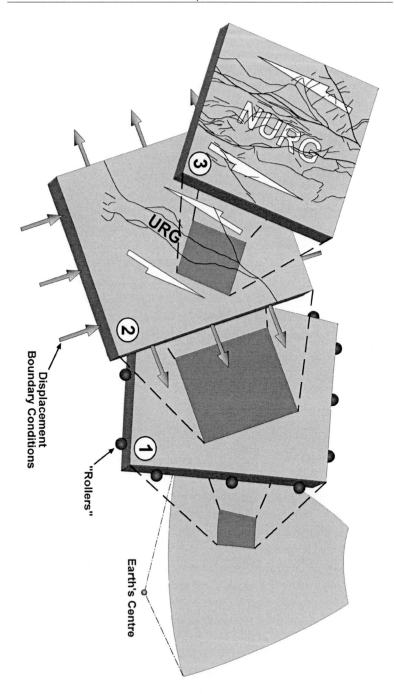

7.3 Model calibration and benchmarking

An important part in the current multi-scale modelling approach is the calibration and benchmarking of the modelling results (i.e. prediction of in-situ stress magnitudes) to independent data and measurements. The FE results were calibrated against stress magnitudes derived from in-situ stress measurements from a geothermal site within the central URG (Figure 4.9). The geothermal site of Soultz sous Forêts is located in the centre of the model within the graben unit. In order to assess the quality of the predicted in-situ stress magnitudes for the entire model domain, the results of the calibrated model were compared to in-situ stress data from an additional geothermal site located in the Swabian Alb in the eastern shoulder region (Bad Urach; Figure 4.10). For the simulation of a state of stress, which is in accordance to the calibration and benchmarking data, it is important to emphasise the necessity of the pre-stressing procedure in order to generate equilibrated gravitationally induced stresses. Following this approach, including the effect of the spherical geometry of the Earth and additionally simulating the effect of internal pore fluid pressure and other contributions to the in-situ state of stress (e.g. TERZAGHI, 1923; GOODMAN, 1980; SHEOREY, 1994), in-situ stress magnitudes are predicted by the model that are consistent with the observation derived from a compilation of world-wide stress magnitude data (Figure 3.6). The predicted state of stress would be characterised by significantly underestimated horizontal stress magnitudes (i.e. $k = S_{Hmean}/S_V = 0.33$) if this pre-stressing approach is not applied. In addition, this would result in a significant overestimation of differential stress magnitudes within the model. Such a non-pre-stressed modelling approach leads to the prediction of a radial extensional state of stress (i.e. $S_V >> S_H = S_h$) for almost the entire crust, which is not in accordance with either the observation from the compilation of world-wide stress magnitude data or the in-situ stress measurement data for the URG area.

By combining the pre-stressing approach described with a calibration and benchmarking procedure, it is possible to simulate the absolute state of stress affecting the geological system. This is a significant improvement for geodynamic 3D numerical modelling including gravitational acceleration, since the modelled state of stress not only represents the change in the stress components induced by the imposed boundary conditions but can also be compared directly to approximations of the in-situ state of stress derived from geomechanical and seismological measurements. Therefore, the modelling results can be used to address geomechanical problems, for example the assessment of borehole stability during drilling, the optimisation of the production process in hydrocarbon industry and the investigation of possible fault reactivation and fracture generation beyond the range of earlier procedures.

7.4 Modelling results and discussion

In contrast to previous numerical modelling studies of parts of the European Cenozoic Rift System (ECRIS; e.g. DIRKZWAGER, 2002; WORUM et al., 2004; SCHWARZ AND HENK, 2005) the approach presented in this thesis focuses on the simulation of absolute stress magnitudes to approximate the absolute crustal state of stress within the URG area. This addresses both the kinematic behaviour of the entire URG system as well as the behaviour of local areas of the graben. Since the complexity of the numerous intra-graben faults cannot be implemented due to limitations in resolution in a large-scale model of the entire URG, a multi-scale approach is used in which the sub-models include more complex geometries. In this section the most important results of the global model and the two sub-models are presented and discussed.

7.4.1 URG scale (global model)

The state of stress observed in the URG area can be explained by the use of the described modelling approach including a combination of gravitational pre-stressing and tectonic loading. The distribution of the various tectonic regimes predicted by the model is predominantly related to gravitational potential energy differences induced by the topographies and lateral density variations implemented within the crust. After the tectonic loading the following tectonic regimes are obtained in the study area (see Figure 4.15):

- An overall strike-slip regime predicted for the Rhenish Massif and the marginal regions of the Paris Basin. In this region, the relationship between crustal thickness and relative horizontal stress magnitudes is obvious, such that relatively increased horizontal stress magnitudes correspond to regions characterised by an increased crustal thickness.

- An overall transtensional state of stress is predicted for the northern and central URG. The central segment of the URG shows slightly elevated relative horizontal stress magnitudes due to the overall change in strike of the central segment of the URG with respect to its northern and southern segments.

- An overall transtensional to extensional state of stress is predicted for the southern segment of the URG, the Vosges Mountains, the Rhine Saône Transfer Zone, the Jura Mountains, the Black Forest and the Swabian Alp. These regions are characterised by both elevated surface and Moho topographies.

- An overall transtensional to strike-slip regime is predicted for the Molasse Basin.

The tectonic (i.e. far-field) loading significantly influences the state of stress in central Europe and the URG area (e.g. MÜLLER et al., 1992; GÖLKE & COBLENTZ, 1996). This is clearly evident from the homogeneous orientation of the maximum horizontal stress component observed in this area (Figure 4.7). Plate boundary forces (i.e. ridge push and Alpine compression) result in distinct S_H orientations and transfer additional horizontal stresses into the URG area. As suggested by several previous authors (e.g. ILLIES, 1974A; ILLIES & GREINER, 1976, SCHUMACHER, 2002) this far-field loading condition probably imposes a sinistral shear couple on the graben (Figure 7.2). The URG global model suggests that this sinistral shear couple is acting on a region typified by varying transtension and that fault displacement is occurring on a series of inherited structures, which will reactivate in response to their local loading. Furthermore, the results of the global model suggest that on a regional scale the imposed loads are dissipated differentially by the reactivation of the first order fault segments. This leads to distinct polarity changes in the style of deformation of the graben, for example, the graben master fault switches from EBF to WBF several times (Figure 4.27 and Figure 7.2). The prediction of strain partitioning along the various segments of the URG border faults is in agreement with observations from a compilation of graben bounding fault segments with documented Pleistocene activity (Figure 4.26) and studies based on the interpretation of seismic sections (MAUTHE et al., 1993; DERER et al., 2005). In addition, the study presented in this thesis demonstrates that in the area of the URG induced vertical strain also has an important influence on the recent graben history. The modelled displacement pattern corresponds directly to the spatial variation of induced Moho uplift. This is also clear for the region of the Rhenish Massif for which the present model fails to predict the observed accelerated recent uplift (e.g. MEYER & STETS, 1998; Figure 7.2) due to the non-implementation of a thermal boundary condition.

The derived kinematic model is applied as boundary conditions on local-scale models (i.e. sub-models). This enables the comparison of the model predictions with geological data due to their higher spatial resolution (see section 7.4.2 and section 7.4.3).

Figure 7.2 (next page): Kinematic model of the URG. Far-field loads impose a sinistral shear couple on the graben system. Additionally, the spatial variation of induced Moho uplift induces distinct areas of uplift (+; i.e. Vosges, Black Forest, Odenwald and Rhenish Massif). The largest relative subsidence is induced in the region of the Heidelberger Loch (-). Within this system, pre-existing first order fault structures are reactivated differentially due to their relative orientation to the imposed far-field loads (black lines indicate increased activity; white lines indicate decreased activity).

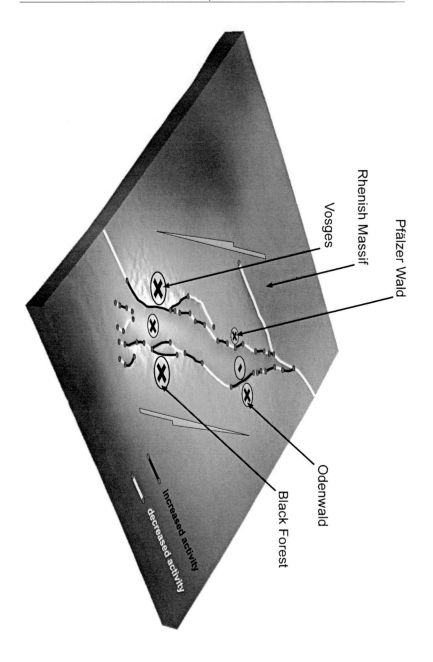

Pfälzer Wald

Rhenish Massif

Vosges

Odenwald

Black Forest

increased activity

decreased activity

7.4.2 Central URG scale (*sub-model A*)

The state of stress in the central segment of the URG is characterised by an overall transtensional tectonic regime. The spatial variation in the dominance of the vertical stress component is primarily induced by the variation in surface topography. S_V is relatively increased in regions with elevated surface topography (e.g. northern Black Forest; Figure 5.5a). Within the graben, a reduced dominance of S_V is predicted due to the reduced surface topography and the sedimentary composition of graben material (i.e. reduced average density). In contrast to the surface topography, the influence of the Moho topography on local variations in the state of stress predicted is negligible since no distinct local variations of the Moho topography are observed. As also indicated by the global model results, the central segment of the URG forms a restraining bend due to the overall change in strike of this segment. This leads to highly variable fault kinematics in this region. In contrast to the northern segment of the URG, the prediction of differential movement along the implemented fault structures indicates that the central segment of the URG behaves more compressionally than the overall URG trend. For this segment of the URG, *sub-model A* predicts that inversion of NNE/ SSW oriented fault structures is likely to occur. The most active (i.e. relatively increased fault slip) fault structure within the *sub-model A* area is predicted to be a segment of the EBF system located in the vicinity of the city of Baden-Baden. For this segment both relatively increased differential displacements (Figure 5.7b) and relatively increased slip tendency (ST) and dilation tendency values (Figure 5.8) are predicted. Comparing the spatial distribution of the recent earthquake activity (1933 to 2002; Figure 5.15) with the ST values calculated for the fault model, the majority of the damaging earthquakes are suggested to be associated with the EBF system, which is characterised by relatively decreased ST values. The comparison of the modelling results with the various reference data shows a good agreement of:

- The modelled stress magnitudes with reference data (e.g. PLENEFISCH & BONJER, 1997; VALLEY & EVANS, 2007).

- The modelled differential displacements within the regional displacement pattern including varying uplift and subsidence of individual fault blocks within the graben (e.g. KUNTZ et al., 1970; ILLIES, 1974A; BARTZ, 1974; Figure 7.3).

- The predicted sense of slip along the modelled faults including the highly variable fault kinematics in this region with geological and geodetic data (e.g. KUNTZ et al., 1970).

- The moment magnitude derived from the modelled fault slip of the simulated 1933 Rastatt earthquake using the pseudo dynamic approach with the seismological observation (AHORNER & SCHNEIDER, 1974).

- The modelled overall strike-slip fault kinematics of the 1933 Rastatt earthquake with the seismological observation (AHORNER & SCHNEIDER, 1974).

Figure 7.3 illustrates the kinematic model derived from the modelling results. The imposed loads induce sinistral reactivation of the central URG. This region is characterised by relatively decreased differential uplift of the graben and shoulder units. Areas of highest differential surface displacement are associated with the EBF. In its central segment, the URG shows a distinct asymmetry, whereas in the area of the western graben the imposed loads are dissipated along several fault structures. In contrast, on the eastern side of the graben the deformation is more localised on the EBF and sub-parallel fault structures, which are oriented unfavourably for reactivation relative to the present-day stress field.

Figure 7.3 (next page): Kinematic model of the central URG. The imposed loads lead to a sinistral reactivation of the region. In this context, only minor differential surface displacements are induced (+, relative uplift; -, relative subsidence). The pseudo fault plane solution indicates the fault structure, which is assumed to be related to the Rastatt 1933 earthquake.

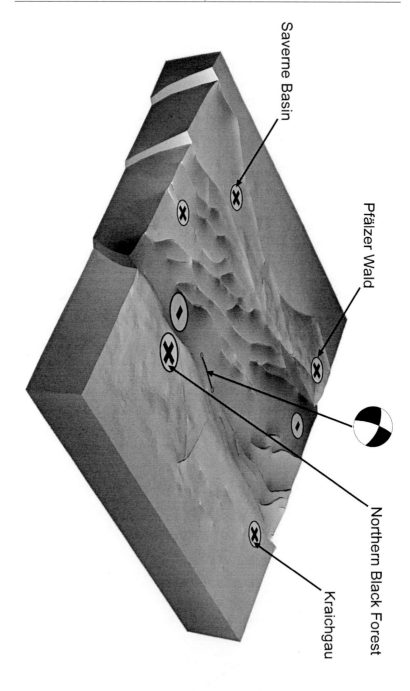

Saverne Basin

Pfälzer Wald

Northern Black Forest

Kraichgau

7.4.3 Northern URG scale (*sub-model B*)

Similarly to the prediction of the global model and *sub-model A*, the results of *sub-model B* suggest that the northern segment of the URG is also characterised by an overall transtensional state of stress (Figure 6.4a). In contrast to the *sub-model A* results, the spatial variability of the dominance of the vertical stress component (S_v) is not primarily induced by the surface topography. In the northern segment, the influence of the more complex crustal composition on the predicted stress components magnitudes is visible. Here, the relative magnitude of S_v is also influenced significantly by the gradient of the Moho topography and local density variations within the crust. Within the graben, a generally reduced dominance of S_v is also predicted for the northern segment of the URG. This is an effect of the reduced surface topography and the different composition of graben material. An exception from this general trend is visible for an approximately 30 km wide corridor to the west of the city of Heidelberg. Here, the predicted relative S_v magnitudes are reduced significantly. As also indicated by the global model results, the northern segment of the URG forms a releasing bend due to the more favourable orientation of this segment relative to the imposed loading. As a result of this transtensional state of stress, this leads to more uniform fault kinematics in this region when compared to *sub-model A*.

In contrast to the central segment of the URG, the prediction of differential movements along the modelled fault structures indicate that the northern segment of the URG, especially within the corridor west of Heidelberg including the Heidelberger Loch depocentre, behaves more extensionally than the overall URG trend. *Sub-model B* predicts the largest relative subsidence of the graben to occur in this part of the URG along the most active fault structure, which is predicted to be a segment of the EBF system located in the vicinity of the city of Heidelberg (i.e. Heidelberger Loch). For this fault segment both relatively increased differential displacements (Figure 6.5) and relatively increased slip tendency and dilation tendency values (Figure 6.9) are predicted. Due to the lack of documented damaging earthquakes in this area, no comparison of a natural distribution with the predicted ST values is possible. One indication that larger earthquakes in the northern segment of the URG are associated with fault structures characterised by low ST values might be the 1952 Worms earthquake (Figure 6.15). Furthermore, the seismically active HTBF is also characterised by reduced ST values. The comparison of the modelling results with the various reference data can be summarised as follows:

- Good agreement of modelled stress magnitudes with reference data (e.g.

PLENEFISCH & BONJER, 1997).

- Generally good agreement of modelled differential displacements with the regional displacement pattern:

  The highest relative surface uplift is predicted for a region corresponding to the Odenwald basement high (Figure 7.4).

  The highest relative surface subsidence is predicted for a region corresponding to the Heidelberger Loch (e.g. BARTZ, 1974; Figure 7.4).

  Both regions are separated by the most active fault structure in the northern URG (i.e. highest differential displacements along the EBF; e.g. BARTZ, 1974; Figure 7.4).

  Within the Mainz Basin the southern border region of the Niersteiner Horst shows largest recent differential uplift and hence, the Niersteiner Horst does not tectonically behave as a horst under the simulated contemporary kinematics (PETERS & VAN BALEN, 2007A; Figure 7.4).

- The model predictions disagree with the geological observation for the region of the Rhenish Massif and parts of the Mainz Basin. No dominant uplift is predicted for this region since the Eifel plume, generally regarded responsible for the uplift, is not implemented in the model (e.g. GARCIA-CASTELLANOS et al., 2000).

- Good agreement of the predicted sense of slip along the modelled faults (e.g. BARTZ, 1974;).

Good agreement of the moment magnitude derived from the modelled fault slip to the simulated possible palaeo-earthquake using the pseudo dynamic approach with the palaeoseismological observation (PETERS et al., 2005).

- Good agreement of the modelled overall extensional fault kinematics of the possible palaeo-earthquake at the WBF with the field observation (PETERS et al., 2005).

Figure 7.4 illustrates the kinematic model derived from the modelling results. The northern segment of the URG is, similar to the central graben segment, characterised by an asymmetric geometry. However, differential uplift is larger in the northern URG. Here, the area of the Odenwald and the Sprendlinger Horst experience increased uplift, whereas within the graben the Heidelberger Loch is characterised by the largest recent subsidence documented in the URG. On the western side of the graben relative vertical motions are distributed over several fault structures. Here, centres of relative uplift are predicted for the Pfälzer Wald and fault blocks located in the SE of the Niersteiner Horst. The distinct tilting of this region predicted by the model is also observed from the study of river terraces (PETERS & VAN BALEN, 2007A). As also seen in the global model results, this model cannot predict the recent uplift of the Rhenish Massif (i.e. Taunus; see section 7.4.1).

Figure 7.4 (next page): Kinematic model of the northern URG. Similar to the central segment, the imposed loads lead to a sinistral reactivation of the region. In contrast, the spatial variation of ongoing mantle uplift induces a distinct locus of uplift in the eastern shoulder region (i.e. Odenwald basement high; +, relative uplift; -, relative subsidence). The pseudo fault plane solution indicates the fault structure, which is related to the possible palaeo-earthquake simulated.

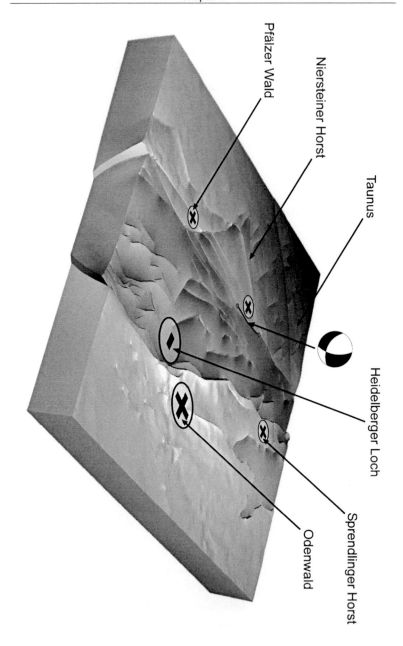

7.5   Integrated discussion of the pseudo dynamic earthquake modelling approach

The combined results of both pseudo dynamic earthquake models can be used to evaluate the quality of the approximated state of stress from the URG models. Table 7.1 shows the scaling relationship of the predicted displacements of the simulated earthquakes relative to the area of the implemented rupture plane.

Earthquake	Rupture area	Reference M_w	Derived M_w
Rastatt	13 km^2	5.035	5.05
Palaeo	144 km^2	6.2	6.17

Table 7.1: Relation of derived moment magnitudes from the individual simulated earthquakes with the assumed rupture area.

The modelling results suggest that, given a valid approximation of the in-situ state of stress and the area of the rupture plane, an earthquake can be adequately simulated with FEm using the pseudo dynamic approach described in section 5.5.4 and section 6.5.4. The modelling results show that the empirical earthquake scaling relationships derived from seismological observation (e.g. KANAMORI & ANDERSON, 1975; WELLS & COPPERSMITH, 1994; GRÜNTHAL & WAHLSTRÖM, 2003; STEIN & WYSESSION 2003) can be reproduced using the continuum mechanics approach of the finite element method. Furthermore, the results indicate that the contact algorithm used in ABAQUS™ is adequate to approximate the displacement that occurs along the modelled rupture plane when a pseudo dynamic approach is used to simulate the change in stress induced by an earthquake. For the determination of the stress change induced by an earthquake, an initially stress free modelling approach is commonly chosen (e.g. KING et al., 1994). Using such an approach, displacement boundary conditions are applied directly to the nodes adjacent to the contact surface (i.e. rupture surface). The advantage of the pseudo dynamic approach used in this thesis is that, in contrast to the more commonly used approach, shear stresses along the considered rupture plane are dissipated due to the induced displacements and no additional energy is transferred into the system by imposing displacements on the nodes adjacent to the rupture plane. Furthermore, the effect of a simulated earthquake on the in-situ state of stress can be investigated. This enables the comparison of the pseudo dynamically obtained parameters (e.g. ΔCFS, $\Delta\tau_{max}$) with parameters obtained from the static loading of the model (e.g. ST, DT). It is important to note that when considering only the ΔCFS parameter, identical magnitudes are predicted by both a stress free and a pre-stressed modelling approach. Figure 7.5 compares the obtained ΔCFS magnitudes using both approaches (see also section 2.3.3). It is commonly argued that ΔCFS alone is adequate to model earthquake interaction. The more complete analysis of a wide range of parameters presented herein clearly shows that this is due to the limitations of the stress free method, rather than the real situation.

To further evaluate the modelling results, the predicted displacement vectors of both earthquakes simulated can be compared to the causative stress field derived from the modelling results at each data point in the model. For this purpose, the kinematic indicators (i.e. displacement vectors) obtained can be analysed using different approaches (e.g. NDA, SPANG, 1972 or direct inversion, ANGELIER & GOGUEL, 1979). Figure 7.6 shows the results of the single event NDA calculation for both simulated earthquakes.

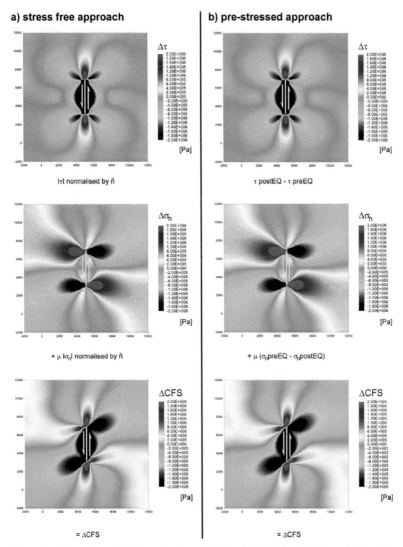

Figure 7.5: Comparison of ΔCFS results for an infinite set of left-lateral faults parallel to a left-lateral master fault from two different modelling approaches. For both models identical slip magnitudes and coefficients of friction (μ) are assumed. a) ΔCFS calculated following the commonly used approach for an initially stress free model (e.g. KING et al., 1994). b) ΔCFS calculated for a gravitationally and tectonically loaded model.

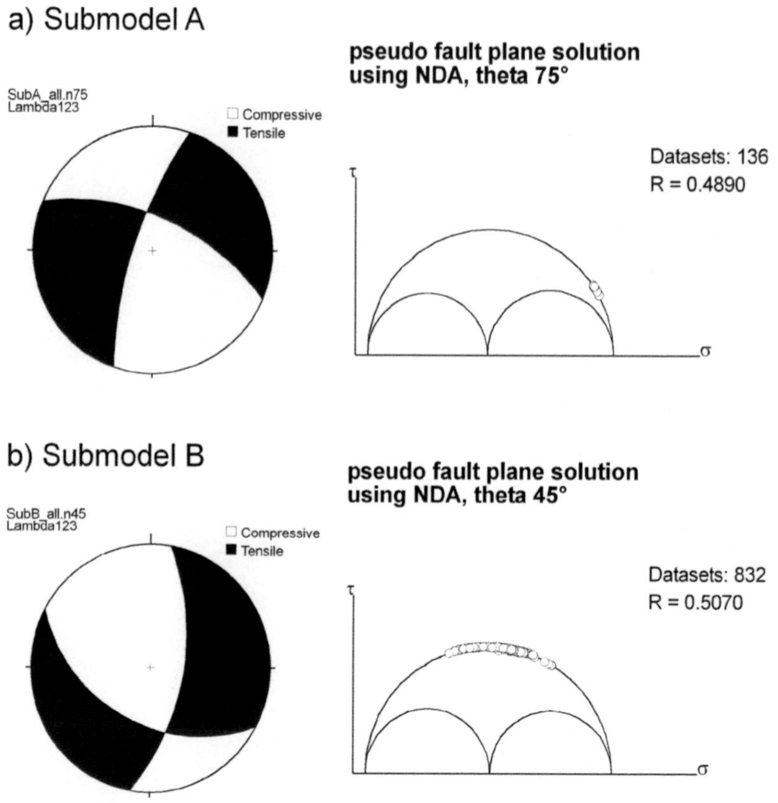

a) Submodel A

**pseudo fault plane solution
using NDA, theta 75°**

SubA_all.n75
Lambda123

☐ Compressive
■ Tensile

Datasets: 136
R = 0.4890

b) Submodel B

**pseudo fault plane solution
using NDA, theta 45°**

SubB_all.n45
Lambda123

☐ Compressive
■ Tensile

Datasets: 832
R = 0.5070

Figure 7.6: Calculated single event pseudo fault plane solution for both, the Rastatt and the postulated palaeo earthquake modelled using the NDA method. a) Results of the NDA calculation for the Rastatt earthquake indicate a pure strike-slip mechanism with only a minor extensional component (R ~ 0.5; θ = 75°; see section 5.5.4). b) Results of the NDA calculation for the assumed palaeo-earthquake indicate a pure extensional mechanism with only a minor strike-slip component (R ~ 0.5; θ = 45°; see section 6.5.4).

In contrast to the single event NDA analysis, the orientation and relative magnitude of the causative stress field can be approximated from the analysis of displacement vectors obtained from multiple events using the direct inversion approach of ANGELIER & GOGUEL (1979). Figure 7.7 shows the results of the direct inversion for the combined simulated earthquakes.

The calculated pseudo fault plane solution obtained using the direct inversion method after ANGELIER & GOGUEL (1979) shows the consistency of the obtained modelling data from the pseudo dynamic and the static approach. Both results suggest an overall transtensional state of stress for the considered seismogenic layer in the northern and central URG.

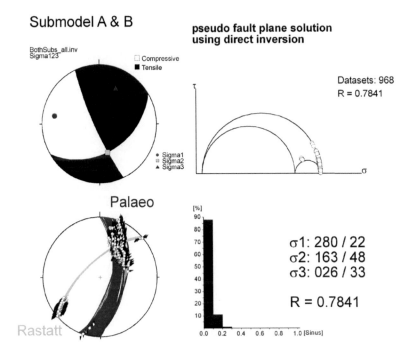

Figure 7.7: Calculated combined event pseudo fault plane solution for both earthquakes modelled using the direct inversion method. The results of the direct inversion reveal a transtensional state of stress (R = 0.7841) under which both, the Rastatt earthquake (strike-slip faulting mechanism; green) and the palaeo-earthquake (extensional faulting mechanism; blue) occurred. The fluctuation histogram shows the minor error of the solution.

7.6 Contribution of 3D finite element modelling to earthquake hazard assessment

In this thesis the parameter of slip tendency (ST; MORRIS et al., 1996) is tested to see whether it is appropriate for the identification of possible sources for damaging earthquakes in the northern and central URG area. The URG area is presently characterised by low to moderate intra-plate seismicity. Few damaging earthquakes (magnitude > 5) have been recorded in historic times (LEYDECKER, 2005A) and the sources for possible future damaging earthquakes are unknown. A large number of tectonic faults occur both within the graben and in the shoulder areas. The geological record shows that during the Quaternary, intra-graben faults and the border faults have been frequently active, possibly causing considerable surface deformation (e.g. ILLIES, 1974B; MONNINGER, 1985). In intra-plate settings, large earthquakes typically occur in recurrence intervals of several thousands to tens of thousands of years. Given the large number of faults and the incomplete seismic catalogue, damaging earthquakes can possibly occur in this highly populated and industrialised region. Present-day seismic hazard assessment of the URG is entirely based on the seismic catalogue of the last 1200 years. For a seismic hazard assessment the contribution of the stress field and tectonic features are essential input parameters. Using the method of 3D finite element modelling, this thesis attempted to define the parameters and the geometric structure necessary to reproduce the contemporary kinematic behaviour of the URG and to quantify the stress field and motion rates of tectonic features of the graben. By defining the contemporary kinematic behaviour and the absolute state of stress, the identification of fault segments which are possible sources for damaging earthquakes might be possible.

A similar approach was previously used to identify possible seismogenic faults in the North German Basin in the framework of the evaluation of the structural safety of the proposed German nuclear repository (LEYDECKER, 2005B). The analysis conducted by LEYDECKER (2005B) was based on a 3D finite element analysis of the state of stress affecting the region of northern Germany (CONNOLLY et al., 2003). In this study the obtained approximation of the in-situ state of stress was used to calculate ST for a pre-existing fault system in a region of the North German Basin. LEYDECKER (2005B) identified fault structures for which the chosen threshold of ST of 0.7 was exceeded and concluded that these faults could possibly be most prone for reactivation in the assumed stress field. This threshold was chosen according to the observation that shear reactivation of a pre-existing fault surface is likely to occur when ST values of 0.7 are reached (MORRIS et al., 1996). The identified fault surfaces were then defined as possible sources for future earthquake events. The estimation of the future earthquake mechanisms is based on the following assumptions (LEYDECKER, 2005B):

- Only fault segments which exceed the critical ST threshold of 0.7 are considered to be potential sources for future earthquake events.

- The location of the maximum ST value predicted along each considered fault surface represents the location of the initiation of slip along the fault (i. e. earthquake hypocentre).

- The maximum rupture length of the earthquake is equal to 40% of the total fault length (LINDENFELD & LEYDECKER, 2004).

- The earthquake mechanism is defined by the average strike and dip of the rupture plane, whereas the rake angle is defined by the orientation of the shear stress on the rupture plane.

- The average moment magnitude (M_W) can be derived from the rupture length using the approach of RUDLOFF & LEYDECKER (2001A and 2001B).

The potential of fault reactivation has been investigated for both URG sub-models using calculations of the dilation tendency (DT) and the slip tendency parameter (ST; see section 5.5.3 and section 6.5.3). Dependant on the assumed coefficient of friction, high slip tendency values are predicted for numerous fault segments in both sub-models. Using the approach of LEYDECKER (2005B) these fault segments could be considered as potential sources for damaging earthquakes (corresponding to possible moment magnitudes of up to 6.8) as long as a minimum segment length requirement is met (of ~12 km).

Using the ST results of both sub-models, faults with orientations of NNW/SSE to NNE/ SSW are more prone to shear reactivation with the NNW/SSE trending faults exhibiting the highest ST values (Figures 5.8b, 5.9, 6.9b). Faults with NNW/SSE orientation occur frequently in the entire URG. In the shoulder areas of the graben, faults strike mainly NW/SE and exhibit low ST values. The comparison of ST results with the geological record shows that faults with high ST values have been frequently reactivated over geological time spans. This activity is documented in numerous studies including the interpretation of seismic profiles, well data and field observation (e.g. STÄBLEIN, 1968; BARTZ, 1974; PFLUG, 1982; HAIMBERGER et al., 2005; PETERS et al., 2005; WIRSING et al., 2007). In contrast, faults with documented historical seismicity are generally predicted to have low ST values (e.g. HTBF or EBF system in the vicinity of Rastatt; Figure 6.9b and Figure 5.8a). The recorded damaging earthquakes occurred on faults with orientations of NNE/SSW to NE/SW. Local segments on these faults have slightly increased ST, but these segment lengths are too short to be relevant for a damaging earthquake (i.e. they are < 12 km).

The pseudo dynamic approach used in this study indicates that, given a valid approximation of the in-situ stress state, the rupture processes during an earthquake event are independent of the ST values along that fault segment. Rather they are primarily determined by the surface area that can be reactivated (see section 7.5.2). This deduction is consistent with the URG observations where seismicity is often associated with faults characterised by an unfavourable orientation relative to the regional stress field (e.g. Rastatt 1933 earthquake). In contrast, favourably oriented faults are characterised by increased relative movements and a lack of recent seismicity (e.g. the EBF in the vicinity of the Heidelberger Loch).

The slip tendency analysis indicates that a large number of fault surfaces within the URG system are favourably oriented for shear reactivation. In general, due to this large number of optimally oriented faults, it may not be possible to accumulate sufficient energy for large earthquakes. Instead, continual releases through many smaller earthquakes (M < 5) or via fault creep (e.g. AHORNER, 1975; BONJER et al., 1984; FRACASSI et al., 2005; PETERS, 2007) maybe the preferred strain accommodation mechanisms. Under this hypothesis, large earthquakes in the URG system are rare and are more likely to be associated with fault structures along which the imposed loading cannot be dissipated by aseismic creep or smaller seismic events. It is therefore concluded that major URG earthquakes occur on non-favourably oriented fault surfaces where the fault plane motion is a result of kinematically imposed relative movements of the two adjacent tectonic entities (i.e. fault blocks).

Using the approach of LEYDECKER (2005B) the localisation of possible future earthquake events has large uncertainties since the predicted ST magnitude is dependant on the effective stress field, the orientation of the fault surfaces and the angle of friction used. Assuming a valid approximation of the in-situ state of stress, the physical (i.e. spatial and mechanical) properties of the individual fault surfaces remain largely unconstrained and are highly influenced by the model resolution. Therefore, using LEYDECKER'S (2005B) ST based method it is not possible to predict the location or timing of future earthquakes, since this method only considers simplified static processes. Furthermore, the method cannot predict the mechanism of movement along the fault surface (stick-slip or creep) due to the simplistic friction laws used. The results of this thesis suggest that, despite these limitations, the ST method allows identification of the location of potential seismic sources. However, the ST results have to be analysed carefully and evaluated against geological data. For the URG area, the modelling results reveal for several fault segments with high fault motions and absence of seismicity at locations of high ST (e.g. EBF in the vicinity of the Heidelberger Loch). This demonstrates that high ST along a fault segment does not necessarily imply a higher potential for large earthquakes for that segment. In case of the URG, it appears that high ST values imply a high potential for a-seismic creeping motions. It is therefore concluded that 3D finite element modelling can contribute with additional input in seismic hazard assessments provided that the results are carefully evaluated and compared against geological data.

7.7   Main Conclusions

The modelling study conducted for this thesis provides a set of technical and regional kinematic conclusions:

Technical conclusions:

- The 3D state of stress in the Earth's crust can be sufficiently approximated using linear elastic material properties provided that a procedure of pre-stressing yields a gravitationally equilibrated initial state of stress which, in addition to the Poisson's Effect, accounts for other contributions to the horizontal stress components (e.g. pore fluid pressure).

- The observed relative increase of the horizontal stress components towards the Earth's surface can be simulated using spherical model geometries including a calibrated artificial compaction layer.

- In order to obtain a valid approximation of the in-situ state of stress, the model has to be calibrated against in-situ stress measurements. In addition, the predictions of the calibrated model have to be compared to independent data sets (e.g. stress indicator data and in-situ stress measurements for other locations). Furthermore, the predicted kinematic behaviour of the geological system should be compared to available geological, geomorphological, geodetical and seismological data.

- The ABAQUS™ *submodelling* technique can be used successfully in geomechanical modelling to transfer obtained displacement fields between different model domains in order to reduce the influence of boundary effects on the area of interest. Moreover, the model resolution can be increased significantly for the area of interest avoiding large, computational expensive model volumes.

- Given a valid approximation of the in-situ state of stress, a pseudo dynamic approach can be used to investigate the stress change induced by an earthquake (i.e. ΔCFS analysis). The advantage of this approach is that the obtained distribution of ΔCFS can be compared to other fault risking parameters such as slip tendency. In addition, the displacement vectors can be used to calculate pseudo fault plane solutions, which can be compared to the regional seismicity for further evaluation of the modelling results.

Kinematic conclusions:

- The overall transtensional crustal state of stress of the URG area is the result of the interaction between far-field loads, possibly resulting from both the Alpine push and Mid-Atlantic ridge push forces, and gravitational potential energy differences induced by crustal thickness and density variations.

- The recent kinematic behaviour of the URG and the adjacent shoulder units can only be reproduced by a model that accounts for vertical strain resulting from ongoing differential Moho uplift possibly induced by lithospheric folding, in addition to the lateral and internal loads.

- Within the present-day stress field, the URG is reactivated as a sinistral shear couple, in which pre-existing fault structures accommodate the imposed loading. In this context, the fault structures are differentially reactivated due to their varying orientation relative to the stress field. This leads to changes in polarity of the graben, with the locus of highest activity located on either the eastern or western graben boundary.

- The results of both sub-models reveal that faulting mechanisms on individual fault surfaces are highly variable with strike-slip and extensional reactivation and minor inversion predicted.

- For both sub-model areas the analyses of the slip tendency and dilation tendency cannot provide constraints on the seismicity of individual fault surfaces. It is however interesting to note that the few seismically active faults in the central and northern URG are characterised by significantly lower slip tendency and dilation tendency values than other faults that show evidence of large displacements over geological timespans.

7.8   Drawbacks and possible improvements

The modelling approach developed during this thesis yields a valid approximation of the observed in-situ state of stress in the URG area. The simulated kinematic behaviour of the URG is in good agreement with independent reference data from geological observation. However, the models presented herein are based on several assumptions and therefore provide room for further improvements.

The modelling results are based on linear elastic material properties. A more appropriate approach for modelling upper crustal stresses would enable the implementation of lithospheric strength profiles and thus the use of an elastic upper crust and visco-elastic lower crust and mantle. In the future, a detailed compilation of the lithospheric strength in Europe could further improve modelling studies of the upper crustal state of stress in the study area (e.g. using data of TESAURO et al., 2007).

The use of visco-elastic material properties requires the lithospheric layers characterised by variable thickness to be in quasi static gravitational equilibrium to maintain their present-day topographies. Therefore, it is necessary to obtain further constraints on the density distribution of the modelling area. Recent investigations on the Earth's gravity field will improve the understanding of the influence of gravitational potential energy differences on the in-situ state of stress (e.g. The European Space Agency's gravity mission GOCE).

A gravitationally equilibrated model including visco-elastic material properties could potentially be used to investigate the effect of lithospheric folding (i.e. buckling) and thermal driven uplift of the Rhenish Massif since the present modelling results indicate that both processes have an important influence on the present-day kinematics of the URG area (e.g. GARCIA-CASTELLANOS et al., 2000; BOURGEOIS et al., 2007).

The modelling series of this thesis is driven by a simplified displacement field derived from GPS measurements and orientation data of the maximum horizontal stress component. A more appropriate approach should provide these boundary conditions from the embedding of the current global model into a plate-scale or quasi plate-scale Europe model. Such a model would enable the study of many local and regional variations (compare to ECKERT, 2007). This Europe-Model, including plate boundary forces as well as a variable thickness of lithospheric layers and a heterogeneous density and strength distribution, can provide insights of the various contributors for the European stress and displacement field. Its results could then be used as boundary constraints for a more detailed modelling study of an arbitrarily chosen region within the model domain.

7.9 Summary and concluding remarks

The modelling approach presented in this thesis is based on an initial gravitational state of stress that is obtained by implementing a variable thickness and density distribution of the individual lithospheric layers using a spherical model geometry and gravitational acceleration towards the Earth's centre. Using linear elastic material properties in a gravitationally accelerated model requires accounting for additional sources of the relative magnitude of the horizontal stress components. Considering only the Poisson's Effect leads to a significant underestimation of the horizontal stress components. Therefore, other contributors to the stress tensor have to be taken into account (e.g. the effect of pore fluid pressure). In order to obtain a more realistic state of stress, the isotropic part of the stress tensor was increased using the concept of gravitational pre-stressing. This gravitationally pre-stressed model is then subjected to tectonic (i.e. far-field) loads and calibrated against in-situ stress measurements, whereas the calibrated modelling results for all model scales provides a realistic 3D approximation of the in-situ state of stress. This yields modelling results, which are consistent with the observed tectonic regimes and enables fault parameters such as slip tendency, dilation tendency and ΔCFS to be integrated and analysed.

A second advantage of the modelling approach developed in this thesis is that it can be applied at various scales ranging from plate-scale to reservoir-scale or smaller. In general, the modelling approach enables the investigation of the effect of plate boundary forces on the individual loading condition of small-scale fault structures, for example in the central URG. The approximated crustal state of stress in a region can be used for a variety of subsequent analyses of economical and societal importance since knowledge of the in-situ stress is vital for the improvement of hazard assessment, reservoir development and reservoir productivity. Since the concept of multi-scale modelling presented in this thesis can be applied on various scales it can be regarded as a multi-purpose approach, which can be applied to geodynamical problems as well as local scale processes. Furthermore, this modelling approach can be further improved to additionally investigate the influence of other geodynamic processes (e.g. buckling).

In summary, the results of the modelling study of this thesis demonstrate that using a simple elastic 3D FE analysis including frictional contact surfaces can be applied successfully to approximate the contemporary state of stress for a geological system at multiple scales. The approach chosen can provide new insights of the kinematics of the system considered such that possible causes for the local surface deformation and local fault behaviour can be investigated. Using this approach a better understanding of the contributions to the in-situ state of stress, regional and local tectonics and the associated local processes can be obtained.

REFERENCES

ABAQUS/Standard User's manual, Abaqus Inc.

Ahorner, L., 1967. Herdmechanismen rheinischer Erdbeben und der seismotektonische Beanspruchungsplan im nordwestlichen Mittel-Europa. Sonderveröffentlichung Geologisches Institut Köln 13: 109-130.

Ahorner, L., 1975. Present-day stress field and seismotectonic block movements along major fault zones in Central Europe. Tectonophysics 29 (1-4): 233-249.

Ahorner, L., 1983. Historical Seismicity and Present-Day Microearthquake Activity of the Rhenish Massif, Central Europe. In: J.H. Illies et al. (Editors), Plateau Uplift: The Rhenish Massif - A Case History, Springer Verlag Heidelberg, pp. 198-221.

Ahorner, L. and Schneider, G., 1974. Herdmechanismen von Erdbeben im Oberrheingraben und in seinen Randgebirgen. In: K. Fuchs and J.H. Illies (Editors), Approaches to Taphrogenesis, Stuttgart, pp. 104-117.

Ahorner, L. and Murawski, H., 1975. Erbebentätigkeit und geologischer Werdegang der Hunsrück-Südrand-Störung. Zeitschrift deutsche geologische Gesellschaft 126: 63-82.

Ahorner, L. and Rosenhauer, W., 1978. Seismic risk evaluation for the Upper Rhine graben and its vicinity. Journal of Geophysics 44: 481-497.

Ahorner, L., Baier, B., Bonjer, K.-P., 1983. General pattern of seismotectonic dislocation and the earthquake-generating field in central Europe between the Alps and the North Sea. In: J.H. Illies et al. (Editors), Plateau Uplift: The Rhenish Massif - A Case History, Springer Verlag Heidelberg, pp. 187-197.

Altenbach, J. and Altenbach, H., 1994. Einführung in die Kontinuumsmechanik, Teubner Studienbücher Mechanik, Stuttgart.

Anderle, H.J., 1968. Die Mächtigkeiten der sandig-kiesigen Sedimente des Quartärs im nördlichen Oberrhein-Graben und der östlichen Untermain-Ebene. Notizblatt des hessischen Landesamtes für Bodenforschung 96: 185-196.

Anderle, H.J., 1987. The evolution of the South Hunsrück and Taunus Borderzone. Tectonophysics 137 (1-4): 101-114.

Anderson, E.M., 1905. The dynamics of Faulting. Edinburgh Geological Society Transactions 8 (3): 387-402.

Anderson, H.-J., 1961. Gliederung und paläogeographische Entwicklung der Chattischen Stufe (Oberoligocän) im Nordseebecken. Meyniana 10: 118-146.

Andre, A.-S., Sausse J. and Lespinasse M., 2001. New approach for the quantification of paleostress magnitudes: application to the Soultz vein system (Rhine graben, France). Tectonophysics 336: 215-231.

Andres, J. and Schad, A., 1959. Seismische Kartierung von Bruchzonen im mittleren und nördlichen Teil des Oberrheintalgrabens und deren Bedeutung für die Ölansammlung. Erdöl und Kohle 12: 323-334.

Angelier, J., Goguel, J., 1979. Sur une méthode simple de détermination des axes principaux es contraintes pour une population de failles. Comptes rendus de l'Académie des sciences, Paris 288: 307-319.

Baier, B. and Wernig, J., 1983. Microearthquake Activity near the Southern Border of the Rhenish Massif. In: J.H. Illies et al. (Editors), Plateau Uplift: The Rhenish Massif - A Case History. Springer-Verlag, Heidelberg, pp. 222-227.

Barth, A., 2002. P-Wellen-Tomographie des oberen Mantels und der Übergangszone unter Eifel und Zentralmassiv. Unpublished diploma thesis, Institute of Geophysics, University of Göttingen, 96 pp.

Barth, A., 2007. Frequency sensitive moment tensor inversion for light to moderate magnitude earthquakes in eastern Africa and derivation of the regional stress field. Unpublished PhD thesis, Institute of Geophysics, University of Karlsruhe, 149 pp. (Barth, 2006 check in text)

Bartz, J., 1974. Die Mächtigkeit des Quartärs im Oberrheingraben. In: Illies, J.H., Fuchs, K. (Editors), Approaches to Taphrogenesis, Stuttgart, pp. 78-87.

Behnke, C., Cloos, H. and Dürbaum, H., 1967. Remarks Concerning the Tectonics of the Upper Rhinegraben. Abhandlungen geologisches Landesamt Baden-Württemberg, 6, The Rhinegraben Progress Report, pp. 3-4.

Behrmann, J.H., Ziegler, P.A., Schmid, S.M., Heck, B., Granet, M., 2005. Special Issue: EUCOR-URGENT Upper Rhine Graben Evolution and Neotectonics. International Journal of Earth Sciences (Geologische Rundschau) 94 (4): 505-778.

Bérard, T. and Cornet, F.H., 2003. Evidence of thermally induced borehole elongation: a case study at Soultz, France. International Journal of Rock Mechanics and Mining Sciences 40: 1121-1140.

Berckhemer, H., 1990. Grundlagen der Geophysik. Wissenschaftliche Buchgesellschaft, Darmstadt.

Berger, J.-P., 1996. Cartes paléogéographiques-palinspastiques du bassin molassique suisse (Oligocéne inférieur - Miocéne moyen). Neues Jahrbuch für Geologie und Paläontologie, Abhandlungen 202 (1): 1-44.

Berger, J.-P., Reichenbacher, B., Becker, D., Grimm, M., Grimm, K.I., Picot, L., Storni, A., Pirkenseer, C. and Schaefer A., 2005. Eocene-Pliocene time scale and stratigraphy of the Upper Rhine Graben (URG) and the Swiss Molasse Basin (SMB). International Journal of Earth Sciences 94: 711-731.

Bodine, J. H., Steckler M.S. and Watts A.B., 1981. Observations of flexure and the rheology of the oceanic lithosphere. Journal of Geophysical Research 86: 3695-370.

Bogaard, P.J.F., Wörner, G., 2003. Petrogenesis of Basanitic to Tholeiitic Volcanic Rocks from the Miocene Vogelsberg, Central Germany. Journal of Petrology 44 (3): 569-602.

Boigk, H., Schöneich, H., 1970. Die Tiefenlage der Permbasis im nördlichen Teil des Oberrheingrabens. In: Illies, J.H., Mueller, S. (Editors), Graben Problems, Stuttgart, 27, pp. 45-55.

Bonjer, K.-P., 1997. Seismicity pattern and style of seismic faulting at the eastern borderfault of the southern Rhine Graben. Tectonophysics 275: 41-69.

Bonjer, K.-P., Gelbke, C., Gilg, B., Rouland, D., Mayer-Rosa, D. and Massinon, B., 1984. Seismicity and dynamics of the Upper Rhinegraben. Journal of Geophysics 55: 1-12.

Bosum, W. and Ullrich, H.J., 1970. Die Flurmagnetometermessung des Oberrheingrabens und ihre Interpretation. Geologische Rundschau 59: 83-106.

Bott, M.H.P., 1959. The mechanisms of oblique slip faulting. Geological Magazine 96: 109-117.

Bott, M.H.P., 1991. Ridge push and associated plate interior stress in normal and hot spot regions. Tectonophysics 200: 17-23.

Bourgeois, O., Ford, M., Diraison, M., Le Carlier de Veslud, Cd., Gerbault, M., Pik, R., Ruby, N., Bonnet, S., 2007. Separation of rifting and lithospheric folding signatures in the NW-Alpine foreland. International Journal of Earth Sciences (Geologische Rundschau) 96: 1003-1031.

Brace, W.F. and Kohlstedt, D.L., 1980. Limits on lithospheric stress imposed by laboratory experiments. Journal of Geophysical Research 85 (B11): 6248-6252.

Breyer, F. and Dohr, G., 1967. Bemerkungen zur Stratigraphie und Tektonik des Rheintal-Grabens zwischen Karlsruhe und Offenburg. Abhandlungen geologisches Landesamt Baden-Württemberg, 6, The Rhinegraben Progress Report, pp. 42-43.

Brudy, M., Zoback, M.D., Fuchs, K., Rummel, F., Baumgärtner, J., 1997. Estimation of the complete stress tensor to 8 km depth in the KTB scientific drill holes: Implications for crustal strength, Journal of Geophysical Research 102: 18,453-18,475.

Brun, J.P., 1999. Narrow rifts versus wide rifts: inferences for the mechanics of rifting from laboratory experiments. Philosophical Transactions of the Royal Society A: Mathematical, Physical and Engineering Sciences, 357 (1753): 695-712.

Brun, J., Wenzel, F., ECORS-DEKORP team, 1991. Crustal-scale structure of the southern Rhinegraben from ECORS-DEKORP seismic reflection data. Geology 19: 758-762.

Brun, J., Gutscher, M.-A., ECORS-DEKORP team, 1992. Deep crustal structure of the Rhine Graben from DEKORP-ECORS seismic reflection data: a summary. Tectonophysics 208: 139-147.

Buchmann, T.J. and Connolly, P.T., 2007. Contemporary kinematics of the Upper Rhine Graben: a 3D finite element approach. Global and Planetary Change 58: 287-309.

Byerlee, J., 1978. Friction of Rocks. Pure and Applied Geophysics 116 (4-5): 615-626.

Cianetti, S., Giunchi, C., Cocco, M., 2005. Three-dimensional finite element modeling of stress interaction: an application to Landers and Hector Mine fault systems. Journal of Geophysical Research 110: B05S17. (Cianetti et al., 2004 check in text)

Cloetingh, S.A.P.L., Wortel R.J.R. and Vlaar N.J., 1982. Evolution of passive continental margins and initiation of subduction zones. Nature 297: 139-142.

Cloetingh, S.A.P.L., Burov, E. and Poliakov, A., 1999. Lithospheric folding: Primary response to compression? (from central Asia to Paris Basin). Tectonics 18: 1064-1083.

Cloetingh, S.A.P.L. and Cornu, T.G.M. (Editors), 2005. Neotectonics and Quaternary fault-reactivation in Europe's intraplate lithosphere. Quaternary Science Reviews 24 (3-4): 235-508.

Cloetingh, S.A.P.L., Cornu, T.G.M., Ziegler, P.A. and Beekman, F., 2006. Neotectonics and intraplate continental topography of the northern Alpine Foreland. EarthScience Reviews 74 (3-4): 127-196.

Cloetingh, S.A.P.L., Bogaard, P., Ziegler, P.A., Wenzel, F., Heidbach, O., Spakman, W., Thybo, H., Jones, A., Bunge, H.P., Andriessen, P., Burov, E., Matenco, L., Bada, G., Peters, G., van Balen, R.T., Facenna, C., Carbonell, R., Gallard, J., Stephenson, R., Artemieva, I., Sliaupa, S., Soesso, A., Onken, O., Ben-Avraham, Z., Friederich, A., Mosar, J. and TOPO-EUROPE Working Group,

2007. 4-D topographic evolution of the intra-plate regions of Europe: a multidisciplinary approach linking geology, geophysics and geotechnology. Global Planetary Change 58: 1-118.

Cloos, H., 1939. Hebung - Spaltung - Vulkanismus. Geologische Rundschau 30: 401-527.

Connolly, P.T., 1996. Prediction of fluid pathways and secondary structures associated with dilational jogs. Unpublished PhD thesis, Imperial College, London, 213 pp.

Connolly, P. T., Gölke, M., Bässler, H., Fleckenstein, P., Hettel, S., Lindenfeld, M., Schindler, A., Theune, U. and Wenzel, F., 2003. Finite Elemente Modellrechnungen zur Erklärung der Auffächerung der größten horizontalen Hauptspannungsrichtung in Norddeutschland - Endbericht -. Institut für Geophysik der Universität Karlsruhe, 163 pp.

Cornet F.H. and Jones R., 1994. Field evidence on the orientation of forced water flow with respect to the regional principal stress directions. Panel discussion. In: P.P. Nelson and S.E. Laubach (Editors), Rock Mechanics. Models and Measurements. Balkema, Rotterdam, pp. 61.71.

Cornet, F.H., Bérard, T. and Bourouis, S., 2007. How close to failure is a granite rock mass at a 5 km depth? International Journal of Rock Mechanics and Mining Sciences 44: 47-66.

Cornu, T.G.M. and Bertrand, G., 2005a. Backward modeling of the rifting kinematics in the Upper Rhine Graben: insights from an elastic-perfect contact law on the restoration of a multi-bloc domain. International Journal of Earth Sciences (Geologische Rundschau) 94: 751-757.

Cornu, T.G.M. and Bertrand, G., 2005b. Numerical backward and foreward modelling of the southern Upper Rhine Graben (France-Germany border): new insights on tectonic evolution of intracontinental rifts. Quaternary Science Reviews 24 (3-4): 353-361.

Cushing, M., Lemeille, F., Cotton, F., Grellet, B., Audru, J.-C. and Renardy, F., 2000. Paleo-earthquakes investigations in the Upper Rhine Graben in the framework of the PALEOSIS project. pp. 39-43.

Davis, R.O., and A.P.S. Selvadurai, 1996. Elasticity and Geomechanics, Cambridge University Press, Cambridge.

Delouis, B., Haessler, H., Cisternas, A. and Rivera, L., 1993. Stress tensor determination in France and neighbouring regions. Tectonophysics 221 (3-4): 413-437.

DeMets, C., Gordon, R.G., Argus, D.F. and Stein, S., 1990. Current Plate Motions. Geophysical Journal International 28: 2121-2124.

Demoulin, A., Pissart, A. and Zippelt, K., 1995. Neotectonic activity in and around the southwestern Rhenish shield (West Germany): indications of a levelling comparison. Tectonophysics 249: 203-216.

Derer, C.E., Kosinowski, M., Luterbacher, H.P., Schäfer, A. and Süß, M.P., 2003. Sedimentary response to tectonics in extensional basins: the Pechelbronn Beds (Late Eocene to early Oligocene) in the northern Upper Rhine Graben, Germany. Geological Society of London Special Publication 208: 55-69.

Derer, C.E., Schumacher, M.E. and Schäfer, A., 2005. The northern Upper Rhine Graben: basin geometry and early syn-rift tectono-sedimentary evolution. International Journal of Earth Sciences (Geologische Rundschau) 94: 640-656.

Dèzes, P. and Ziegler, P.A., 2002. Moho depth map of Western and Central Europe, World Wide Web Address: http://www.unibas.ch/eucor-urgent.

Dèzes, P., Schmid, S.M. and Ziegler, P.A., 2004. Evolution of the European Cenozoic Rift System: interaction of the Alpine and Pyrenean orogens with their foreland lithosphere. Tectonophysics 389: 1-33.

Dèzes, P., Schmid, S.M., Ziegler, P.A., 2005. Reply to comments by L. Michon and O. Merle on "Evolution of the European Cenozoic Rift System: interaction of the Alpine and Pyrenean orogens with their foreland lithosphere by P. Dèzes, S.M. Schmid and P.A. Ziegler, Tectonophysics 389 (2004) 1-33. Tectonophysics 401, 257-262.

Dirkzwager, J.B., 2002. Tectonic modelling of vertical motion and its near surface expression in The Netherlands. Unpublished PhD thesis, Vrije Universiteit, Amsterdam, 156 pp.

Doebl, F. and Olbrecht, W., 1974. An Isobath of the Tertiary Base in the Rhinegraben. In: K. Fuchs and J.H. Illies (Editors), Approaches to Taphogenesis, Stuttgart, pp. 71-72.

Duringer, P., 1988. Les conglomérats des bordures du rift cénozoïque rhénan. Dynamique sédimentaire et contrôle climatique. Unpublished PhD thesis, Univ. Louis Pasteur, Strasbourg (France), 278 pp.

Eckert, A., 2007. 3D multi-scale finite element analysis of the crustal state of stress in the Western US and the Eastern California Shear Zone, and implications for stress - fluid flow interactions for the Coso Geothermal Field. Unpublished PhD thesis, Institute of Geophysics, University of Karlsruhe, 116 pp.

Eckert, A., Connolly, P.T., 2004. 2D Finite Element Modelling of Regional and Local Fracture Networks in the Eastern California Shear Zone and Coso Range, California USA. Geothermal Resources Council Transaction 28: 643-648.

Edel, J. B. and Weber, K., 1995. Cadomian terranes, wrench faulting and thrusting in the central Europe Variscides: geophysical and geological evidence. International Journal of Earth Sciences 84 (2): 412-432

Eisbacher, G.H., 1996. Einführung in die Tektonik. Spektrum Akademischer Verlag, Thieme, Enke, Stuttgart, 374 pp.

Engelder T., 1985. Loading paths to joint propagation during a tectonic cycle: an example from the Appalachian Plateau, USA. Journal of Structural Geology 7: 459-476.

Engelder, T., 1993. Stress regimes in the lithosphere, Princeton University Press, Princeton.

ENTEC, Homepage, http://www.geo.vu.nl/users/entec.

EUCOR-URGENT, Homepage: http://comp1.geol.unibas.ch/

Ferrill, D.A., Winterle, J., Wittmeyer, G., Sims, D., Colton, S., Armstrong, A. and Morris, A.P., 1999. Stressed Rock Strains Groundwater at Yucca Mountain, Nevada. GSA Today 9 (5): 1-8.

Fracassi, U., Nivière, B. and Winter, T., 2005. First appraisal to define prospective seismogenic sources from historical earthquake damages in southern Upper Rhine Graben. Quaternary Science Reviews 24: 401-423.

Franke, W.R., 1989. Tectonostratigraphic units in the Variscan Belt of Europe. Geological Society of America Bulletin 230: 67-90.

Fuchs, K., von Gehlen, K., Mälzer, H., Murawski, H. and Semmel, A. (Editors), 1983. Plateau Uplift, the Rhenish Shield - a Case History, Springer-Verlag, Berlin, 411 pp.

Garcia-Castellanos, D., Cloetingh, S.A.P.L. and Van Balen, R.T., 2000. Modelling the Middle Pleistocene uplift in the Ardennes-Rhenish Massif: thermo-mechanical weakening under the Eifel? Global and Planetary Change 27: 39-52.

Gaucher, E., Cornet, F.H. and Bernard, P., 1998. Induced seismicity for fracture identification and stress determination. In: EUROCK 98, Proceedings of the SPE-ISPM Rock Mechanics in Petroleum Engineering Conference 1: 545-554.

Gebhardt, H., 2003. Palaeobiogeography of Late Oligocene to Early Miocene Central European Ostracoda and Foraminifera: progressive isolation of the Mainz Basin, northern Upper Rhine Graben and Hanau Basin/Wetterau. Palaeogeography, Palaeoclimatology, Palaeoecology 201: 343-354.

Genter, A., Tenzer, H., 1995. Geological monitoring of GPK2 HDR borehole, 1420-3880 m (Soultz-sous-Forêts, France). BRGM report R 38629, 46 pp.

Gölke, M., 1996. Patterns of stress in sedimentary basins and the dynamics of pull-apart basin formation. Unpublished PhD thesis, Vrije Universiteit Amsterdam, The Netherlands, 167 pp.

Gölke, M. and Coblentz, D., 1996. Origins of the European regional stress field. Tectonophysics 266: 11-24.

Goodman, R. E., 1980. Introduction to Rock Mechanics. Wiley, New York, 492 pp.

Gramann, F., 1966. Das Oligozän der Hessischen Senke als Bindeglied zwischen Nordseebecken und Rheintalgraben. Zeitschrift der Deutschen Geologischen Gesellschaft 115: 497-514.

Greiner, G., Illies, J. H., 1977. Central Europe: Active or residual tectonic stresses. Pure and Applied Geophysics 115: 11-26.

Grimm, K.I., 1994. Paläoökologie, Paläogeographie und Stratigraphie im Mainzer Becken, im Oberrheingraben, in der Hessischen Senke und in der Leipziger Bucht während des mittleren Rupeltons (Fischschiefer/ Rupelium/ Unteroligozän). Mitteilungen der Pollichia 81: 7-193.

Groshong, R.H., 1996. Construction and validation of extensional cross sections using lost area and strain, with application to the Rhine Graben. In: P.G. Buchanan and D.A. Nieuwland (Editors), Modern developments in structural interpretation. Geological Society of London Special Publication, pp 79-87.

Grünthal, G. and Stromeyer, D., 1992. The recent crustal stress field in Central Europe: trajectories and finite element modeling. Journal of Geophysical Research 97 (B8): 11,805-11,820.

Grünthal, G. and Wahlström, R., 2003. An Mw based earthquake catalogue for central, northern and northwestern Europe using a hierarchy of magnitude conversions. Journal of Seismology 7: 507-531.

GÜK500, 1998. Geologische Übersichtskarte von Baden-Württemberg 1:500000. Landesamt für Geologie, Rohstoffe und Bergbau Baden-Württemberg, Freiburg i.Br.

Haimberger, R., Hoppe, A. and Schäfer, A., 2005. High-resolution seismic survey on the Rhine River in the northern Upper Rhine Graben. International Journal of Earth Sciences (Geologische Rundschau) 94 (4): 657-668.

Hancock, P. L. and Engelder, T., 1989. Neotectonic joints. GSA Bulletin 101 (10): 1197-1208

Hardebeck J.L., Nazareth J.J. and Haukson E., 1998. The static stress change triggering model: constraints from two Southern California aftershock sequences. Science 291: 2101-2102.

Harris R., 1998. Introduction to special section: Stress triggers, stress shadows, and implications for seismic hazard. Journal of Geophysical Research 103: 24347-24358.

Heidbach, O., Fuchs, K., Müller, B., Reinecker, J., Sperner, B., Tingay, M. and Wenzel, F., (Editors), 2007. The World Stress Map - Release 2005, Commission for the Geological Map of the World, Paris.

Heidbach, O., Hergert, T., Buchmann, T., Eckert, A., Peters, G. and Müller, B., submitted. How to define an initial in-situ stress state for 3D numerical models? International Journal of Rock Mechanics and Mining Sciences.

Heinemann, B., Troschke, B. and Tenzer, H., 1992. Hydraulic investigation and stress evaluations at the HDR test site Urach III, Germany, Geothermal Resources Council Transactions, Davis, CA, pp. 425-431.

Heinemann, B., 1994. Results of scientific investigations at the HDR test site Soultz-sous-Forêts, Alsace (1987-1992). SOCOMINE report, 126 pp.

Helm J. A., 1996. Aléa sismique naturel et sismicité induite du projet géothermique européen RCS de Soultz-sous-Forêts. Unpublished PhD thesis, University Louis Pasteur-EOPGS, Strasbourg.

Henk, A., 1993. Subsidenz und Tektonik des Saar-Nahe-Beckens (SW-Deutschland). Geologische Rundschau 82: 3-19.

Hickman, S., and Zoback M. D., 2004. Stress orientations and magnitudes in the SAFOD Pilot Hole. Geophysical Research Letters 31: L15S12, doi: 10.1029/2004/GL020043.

Hiller, W., 1934. Der Herd des Rastatter Bebens am 8. Februar 1933. Gerlands Beiträge zur Geophysik 41: 170-180.

Hillis, R.R. and Nelson, E.J., 2005. In situ stresses in the North Sea and their applications: petroleum geomechanics from exploration to development. In: A.G. Doré and B.A. Vining (Editors), Petroleum Geology: North-West Europe and Global Perspectives - Proceedings of the 6th Petroleum Geology Conference, pp. 551-564.

Hinsken, S., Ustaszewski, K. and Wetzel, A.,2007. Graben width controlling syn-rift sedimentation: the Palaeogene southern Upper Rhine Graben as an example. International Journal of Earth Sciences (Geologische Rundschau) 96/6: 979-1002.

Hinzen, K.G., 2003. Stress field in the Northern Rhine area, Central Europe, from earthquake fault plane solutions. Tectonophysics 377: 325-356.

Hubert-Ferrari, A., Barka, A., Jacques, E., Nalbant, S.S., Meyer, B., Armijo, R., Tapponier, P. and King, G.C.P., 2000. Seismic hazard in the Marmara Sea region following the 17 August 1999 Izmit earthquake. Nature 404: 269-273.

Illies, J.H., 1965. Bauplan und Baugeschichte des Oberrheingrabens. Ein Beitrag zum Upper Mantle Project. Oberrheinische geologische Abhandlungen 14: 1-54.

Illies, J.H., 1967. Development and Tectonic Pattern of the Rhinegraben. Abhandlungen geologisches Landesamt Baden-Württemberg, 6, The Rhinegraben Progress Report, pp. 7-9.

Illies, J.H., 1974a. Taphrogenesis and Plate Tectonics. In: K. Fuchs and J.H. Illies (Editors), Approaches to Taphrogenesis. Schweizerbart'sche Verlagsbuchhandlung, Stuttgart, pp. 433-460. (check for a or b in text)

Illies, J.H., 1974b. Taphrogenesis, Introductory Remarks. In: K. Fuchs and J.H. Illies (Editors), Approaches to Taphrogenesis. Schweizerbart'sche Verlagsbuchhandlung, Stuttgart, pp. 1-13. (check for a or b in text)

Illies, J.H., 1975. Intraplate tectonics in stable Europe as related to plate tectonics in the Alpine system. Geologische Rundschau 64: 677-699.

Illies, J.H., 1977. Ancient and recent rifting in the Rhinegraben. Geologie en Mijnbouw 56: 329-350.

Illies, J.H., 1981. Mechanisms of Graben Formation. Tectonophysics 73: 249-266.

Illies, J.H. and Fuchs, K., 1974. Approaches to Taphrogenesis. Approaches to Taphrogenesis, 8. E. Schweizerbart`sche Verlagsbuchhandlung (Nägele u. Obermiller), Stuttgart, 460 pp.

Illies, J.H. and Greiner, G., 1976. Regionales stress-Feld und Neotektonik in Mitteleuropa. Oberrheinische geologische Abhandlungen 25: 1-40.

Illies, J.H. and Greiner, G., 1978. Rhinegraben and the Alpine system. Bulletin Geological Society of America 89: 770-782.

Illies, J.H. and Greiner, G., 1979. Holocene movements and state of stress in the Rhinegraben rift system. Tectonophysics 52: 349-359.

Jaeger, J.C., Cook, N.G.W., 1979. Fundamentals of rock mechanics. Methuen and Co., Ltd., London, 3rd edition, 513 pp.

Jaeger, J.C., Cook N.G.W. and Zimmerman R.W., 2007. Fundamentals of Rock Mechanics, 4th edition, Blackwell Publishing, 475 pp.

Jarosinski, M., Beekman, F., Bada, G. and Cloetingh, S.A.P.L., 2006. Redistribution of recent collision push and ridge push in Central Europe: insights from FEM modelling. Geophysical Journal International 167: 860-880.

Jung R., 1991. Hydraulic fracturing and hydraulic testing in the granitic section of borehole GPK1, Soultz-sous-Forêts. Geothermal Science and Technology 3: 149-198.

Kadolsky, D., 1988. Stratigraphy and mollusc faunas of "Landschneckenkalk" and "Cerithienschichten" in the Mainz Basin (Late Oligocene to Early Miocene?) - stratigraphical, paleogeographical and paleoecological results. Geologisches Jahrbuch Reihe A 110: 69-113.

Kälin, D., 1997. Litho- und Biostratigraphie der mittel- bis obermiozänen Bois de Raube-Formation (Nordwestschweiz). Ecologae Geologicae Helvetiae 90: 97-114.

Kanamori, H. K. and Anderson D. L., 1975. Theoretical basis of some empirical relations in seismology, Bulletin Geological Society of America 65: 1073-1095.

Keller, J., Kraml, M. and Henjes-Kunst, F., 2002. 40Ar/39Ar single crystal laser dating of early volcanism in the Upper Rhine Graben and tectonic implications. Schweizer Mineralogisch Petrographische Mitteilungen 82: X1-X10.

King G., Stein R. and Lin J., 1994. Static stress changes and triggering of earthquakes. Bulletin of the Seismological Society of America 84: 935-953.

Klee, G., and Rummel F., 1993. Hydrofrac stress data for the European HDR research project test site Soultz-sous-forêts. International Journal of Rock Mechanics and Mining Sciences and Geomechanics Abstracts 30: 973-976

Klee, G. and Rummel, F., 1999. Stress regime in the Rhinegraben basement and in the surrounding tectonic units. Bulletin d'Hydrogéologie 17: 135-142.

Krantz, R.W., 1991. Measurements of friction coefficients and cohesion for faulting and fault reactivation in laboratory models using sand and sand mixtures. Tectonophysics 188: 203-207.

Kreemer, C., Holt, W.E., 2001. A No-net-rotation Model of Present Day Surface Motion. Geophysical Research Letters 28: 4407-4410.

Kuntz, E., Mälzer, H. and Schick R., 1970. Relative Krustenbewegungen im Bereich des Oberrheingrabens. In: Illies, J.H., Mueller, S. (Editors), Graben Problems, Stuttgart, 27, pp. 170-177.

Larroque, J.M., Etchecopar, A. and Philip, H., 1987. Evidence for the permutation of stresses s1 and s2 in the Alpine foreland: the example of the Rhine graben. Tectonophysics 144: 315-322.

Laubscher, H., 1986. The eastern Jura: relations between thin-skinned and basement tectonics, local and regional. Geologische Rundschau 73: 535-553

Laubscher, H., 1992. Jura kinematics and the Molasse Basin. Ecologae Geologicae Helvetiae 67: 121-133.

Laubscher, H., 2001. Plate interactions at the southern end of the Rhine graben. Tectonophysics 343 (1-2): 1-19.

Leydecker, G., 2005a. Erdbebenkatalog für die Bundesrepublik Deutschland mit Randgebieten für die Jahre 800 - 2004, Bundesanstalt für Geowissenschaften und Rohstoffe BGR, Hannover. (check for a or b and Leydeker in text)

Leydecker, G., 2005b. Projekt Gorleben - Standsicherheit Nachbetriebsphase: Seismische Gefährdung. Teilprojekt Ingenieurseismologie - Abschlussbericht -, Bundesanstalt für Geowissenschaften und Rohstoffe, Hannover. pp. 86.

Lin, J. and Stein, R. S., 2004. Stress triggering in thrust and subduction earthquakes, and stress interaction between the southern San Andreas and nearby thrust and strike-slip faults. Journal of Geophysical Research 109: B02303

Lindenfeld, M. and Leydecker, G., 2004. Bestimmung des Verhältnisses Bruchlänge zu Störungslänge sowie Ergebnisse der Gleittendenzanalyse entlang neotektonischer Störungen in Norddeutschland. Bundesanstalt für Geowissenschaften und Rohstoffe, Hannover. pp. 100.

Link, K., Rahn, M., Keller, J., Stuart, F., 2004. Ft and (U-Th/He) analyses of the Upper Rhine Graben rift flanks - the thermo-tectonic evolution from Cretaceous to recent times. Joint Earth Science Meeting SGF-GV, Strasbourg 20.-25.09. 2004. Abstract Number RSTGV-A-00232.

Lister, C. R. B., 1975. Gravitational drive on oceanic plates caused by thermal. Contraction. Nature 257: 663-665

Lopes Cardozo, G.G.O., 2004. 3-D geophysical imaging and tectonic modelling of the active tectonics of the Upper Rhine Graben Region. Unpublished PhD thesis, Vrije Universiteit Amsterdam, The Netherlands, 163 pp.

Lopes Cardozo, G.G.O., Edel, J.B. and Granet, M., 2005. Detection of active crustal structures in the Upper Rhine Graben using local earthquake tomography, gravimetry and reflection seismics. Quaternary Science Reviews 24: 337-344.

Lutz, M., Cleintuar, M., 1999. Geological results of hydrocarbon exploration campaign in the southern Upper Rhine Graben. Bulletin für Angewandte Geologie 4: 3-80.

Mastin, L.G. and Heinemann, B., 1988. Evaluation of the caliper and televiewer data from the Soultz well between 1400 m and 2000 m. Internal report to Geophysics Institute, University Karlsruhe.

Mauthe, G., Brink, H.J., Burri, P., 1993. Kohlenwasserstoffvorkommen und -potential im deutschen Teil des Oberrheingrabens. Bulletin der Vereinigung Schweizerischer Petroleum-Geologen und Ingenieure 60 (137): 15-29.

Mayer-Rosa, D. and Cadiot, B., 1979. A review of the 1356 Basel earthquake: basic data. Tectonophysics 53: 325-333.

Mayer-Rosa, D. and Baer, M., 1992. Earthquake Catalogue of Switzerland, Swiss Federal Institute of Technology Zurich, Swiss Seismological Service, Zurich.

McCutchen, W., 1982. Some elements of a theory of in situ stress. Technical Note, International Journal of Rock Mechanics 19: 201-203.

McGarr, A. and Gay, N.C., 1978. State of stress in the earth's crust. Annual Review of Earth and Planetary Science Letters 6: 405-436.

McKenzie D., 1969. Speculations on the consequences and causes of plate motions. Geophysical Journal of the Royal Astronomical Society 18: 1-32.

Meghraoui, M., Delouis, B., Ferry, M., Giardini, D., Huggenberger, P., Spottke, I. and Granet, M., 2001. Active Normal Faulting in the Upper Rhine Graben and Paleoseismic Identification of the 1356 Basel Earthquake. Science 293: 2070-2073.

Meier, L. and Eisbacher, G.H., 1991. Crustel kinematics and deep structure of the northern Rhine graben, Germany. Tectonics 10 (3): 621-630.

Meyer, W. and Stets, J., 1998. Junge Tektonik im Rheinischen Schiefergebirge und ihre Quantifizierung. Zeitschrift der deutschen geologischen Gesellschaft 149: 359-379.

Michon, L., Van Balen, R.T., Merle, O. and Pagnier, H., 2003. The Cenozoic evolution of the Roer Valley Rift System integrated at a European scale. Tectonophysics 367: 101-126.

Michon, L. and Merle, O., 2005. Discussion on "Evolution of the European Cenozoic Rift System: interaction of the Alpine and Pyrenean orogens with their foreland lithosphere" by P. Dèzes, S.M. Schmid and P.A. Ziegler, Tectonophysics 389 (2004) 1-33. Tectonophysics 401: 251-256.

Monninger, R., 1985. Neotektonische Bewegungsmechanismen im mittleren Oberrheingraben. Unpublished PhD thesis, Universität Karlsruhe, Karlsruhe, 219 pp.

Morris, A., Ferrill, D.A. and Henderson, D.B., 1996. Slip-tendency analysis and fault reactivation. Geology 24(3): 275-278.

Müller, B., Zoback, M.L., Fuchs, K., Mastin, L., Gregersen, S., Pavoni, N., Stephansson, O. and Ljunggren, C., 1992. Regional patterns of tectonic stress in Europe. Journal of Geophysical Research 97 (B8): 11783-11803.

Müller, B., V. Wehrle, S. Hettel, B. Sperner, and Fuchs F., 2003. A new method for smoothing oriented data and its application to stress data. In: M. Ameen (Editor) Fracture and In-situ Stress Characterization of Hydrocarbon Reservoirs. Geological Society of London, pp. 107-126.

Murawski, H., 1975. Die Grenzzone Hunsrück / Saar-Nahe-Senke als geologisch-geophysikalisches Problem. Zeitschrift der Deutschen Geologischen Gesellschaft 126: 49-62.

Nagel R., 1994. Das Spannungsfeld in der Geothermiebohrung Soultz-sous-forêts abgeleitet aus vertikalen Strukturen in einer Tiefe von 1.9 bis 3.6 km. Unpublished diploma thesis, Universität Karlsruhe.

Nocquet, J.-M., Calais, E., 2004. Geodetic Measurements of Crustal Deformation in the Western Mediterranean and Europe. Pure and Applied Geophysics 161: 661-681.

Oreskes, N., Shrader-Frechette, K. and Belitz, K., 1994. Verification, validation, and confirmation of numerical models in earth sciences. Science 263: 641-646.

PALEOSIS, 2000. Evaluation of the potential for large earthquakes in regions of present-day low seismicity activity in Europe. Final report, project no. ENV4-CT97-0578, Directorate-General XII for Science, Research and Development, Brussels.

Peters, G., 2007. Active tectonics in the Upper Rhine Graben - Integration of paleoseismology, geomorphology and geomechanical modeling. PhD thesis, Vrije Universiteit Amsterdam, Logos Verlag, 270 pp, online version: http://www.ubvu.vu.nl/

Peters, G., Buchmann, T.J., Connolly, P., van Balen, R., Wenzel, F., Cloetingh, S.A.P.L., 2005. Interplay between tectonic, fluvial and erosional processes along the Western Border Fault of the northern Upper Rhine Graben, Germany. Tectonophysics 406: 39-66.

Peters, G., van Balen, R.T., 2007. Tectonic geomorphology of the northern Upper Rhine Graben, Germany. Global Planetary Change 58: 310-334.

Peters, G., van Balen, R.T., 2007. Pleistocene tectonics inferred from fluvial terraces of the northern Upper Rhine Graben, Germany. Tectonophysics 430: 41-65.

Pflug, R., 1982. Bau und Entwicklung des Oberrheingrabens. Erträge der Forschung, 184. Wissenschaftliche Buchgesellschaft, Darmstadt, 145 pp.

Plaumann, S., 1987. Karte der Bouguer-Anomalien in der Bundesrepublik Deutschland 1:500000. Geologisches Jahrbuch, Reihe E (Heft 40): 3-7.

Plenefisch, T. and Bonjer, K.-P., 1997. The stress field in the Rhine Graben area inferred from earthquake focal mechanisms and estimation of frictional parameters. Tectonophysics 275: 71-97.

Price, N.J, 1966. Fault and joint development in brittle and semi-brittle rock, Pergamon Press, Oxford, 176 pp.

Price, N.J., Cosgrove, J.W., 1990. Analysis of geological structures. Cambridge University press, Cambridge. 502 pp.

Prinz, H. and Schwarz E., 1970. Nivellement und rezente tektonische Bewegungen im nördlichen Oberrheingraben. In: J.H. Illies and S. Mueller (Editors) Graben Problems. Stuttgart, 27, pp. 177-183.

Prodehl, C., Mueller, S., Haak, V., 1995. The European Cenozoic rift system. In: K.H. Olsen (Editor), Continental Rifts: Evolution, Structure, Tectonics. Developments in Geotectonics. New York, Elsevier Science 25, pp. 133-212.

Ramsay, J.G. and Lisle, R.J., 2000. The Techniques of Modern Structural Geology, Volume 3: Applications of Continuum Mechanics in Structural Geology. Academic Press, London, 1061 pp.

Ranalli, G., 1992. Average lithospheric stresses induced by thickness change: A linear approximation. Physics of the Earth and Planetary Interiors 69: 263-269.

Ranalli, G., 1995, Rheology of the earth, Chapman & Hall, London, 413 pp.

Reasenberg, P. A. and Simpson R .W., 1992. Response of regional seismicity to the static stress change produced by the Loma Prieta earthquake, Science 255: 1687-1690.

Reichenbacher, B., 2000. Das brakisch-lakustrine Oligozän und Unter-Miozän im Mainzer Becken und Hanauer Becken: Fischfaunen, Paläoökologie, Biostratigraphie, Paläogeographie. Courier Forschungs-Institut Senckenberg 222: 1-143.

Reinecker, J., Heidbach, O., Tingay, M., Connolly, P., Müller, B., 2004. The 2004 release of the World Stress Map (available online at www.world-stress-map.org).

Ritter J.R.R., Jordan M., Christensen U.R., Achauer U., 2001. A mantle plume below the Eifel volcanic fields, Germany. Earth and Planetary Science Letters 186 (1): 7-14.

Rothausen, K. and Sonne, V., 1984. Mainzer Becken. Sammlung geologischer Führer, 79. Gebrüder Borntraeger, Berlin - Stuttgart, 203 pp.

Rotstein, Y., Edel, J.-B., Gabriel, G., Boulanger, D., Schaming, M. and Munschy, M., 2006. Insight into the structure of the Upper Rhine Graben and its basement from a new compilation of Bouguer Gravity. Tectonophysics 425 (1-4): 55-70.

Rózsa, S., Heck, B., Mayer, B., Seitz, K., Westerhaus, M., Zippelt, K., 2005. Determination of displacements in the upper Rhine graben Area from GPS and leveling data. International Journal of Earth Sciences (Geologische Rundschau) 94 (4): 538-549.

Rudloff, A. and Leydecker, G., 2001a. Ableitung von empirischen Beziehungen zwischen verschiedenen Magnituden-Skalen. BGR, Hannover. pp. 46.

Rudloff, A. and Leydecker, G., 2001b. Empirische Beziehungen zwischen Magnituden und geometrischen Bruchgrößen von Erdbeben. Hannover. pp. 39.

Rudloff, A. and Leydecker, G., 2002. Ableitung von empirischen Beziehungen zwischen der Lokalbebenmagnitude und makroseismischen Parametern - Ergebnisbericht, BGR, Hannover.

Rummel, F., Möhring-Erdmann, G. and Baumgärtner, J., 1986. Stress constraints and hydrofracturing stress data for the continental crust. Pure and Applied Geophysics 124 (4-5): 875-895.

Rummel, F. and Baumgärtner, J., 1991. Hydraulic Fracturing Stress Measurements in the GPK1 Borehole, Soultz-sous-Forêts. Geothermal Science and Technology, 3(1-4): 119-148.

Schmidt-Kittler, N. (Editor), 1987. International Symposium on Mammalian Stratigraphy and Palaeoecology of European Palaeogene, Friedrich Pfeil Verlag, München, 312 pp.

Schneider, G., Schick, R. and Berckhemer, H., 1966. Fault-plane solutions of earthquakes in Baden-Wuerttemberg. Zeitschrift für Geophysik 32: 383-393.

Scholz, C. H., 2000. Evidence for a strong San Andreas fault. Geology 28: 163-166.

Schumacher, M.E., 2002. Upper Rhine Graben: Role of preexisting structures during rift evolution. Tectonics 21 (1): 10.1029/2001TC900022.

Schwarz, E., 1974. Levelling Results at the Northern End of the Rhinegraben. In: K. Fuchs and J.H. Illies (Editors), Approaches to Taphrogenesis, Stuttgart, pp. 261-268.

Schwarz, M., 2006. Evolution und Struktur des Oberrheingrabens quantitative Einblicke mit Hilfe dreidimensionaler thermomechanischer Modellrechnungen. Unpublished PhD thesis, University of Freiburg, Freiburg i. Br., 337 pp.

Schwarz, M. and Henk, A., 2005. Evolution and structure of the Upper Rhine Graben: insights from three-dimensional thermomechanical modelling. International Journal of Earth Sciences (Geologische Rundschau) 94: 732-750.

Schwarz, J., Beinersdorf, S., Golbs, C., Ahorner, L. and Meidow, H., 2006. Erdbebenkatalog für Deutschland und angrenzende Gebiete - erweiterter Ahorner-Katalog (EKDAG2006). Auszug mit Schadenbeben der Intensität > VI, Bauhaus-Universität Weimar, Erdbebenzentrum, Köln/Weimar.

Semmel, A., 1968. Studien über den Verlauf jungpleistozäner Formung in Hessen. Frankfurter geographische Hefte 45: 133 pp.

Sheorey, P.R., 1994. A Theory for In Situ Stresses in Isotropic and Transversely Isotropic Rock. International Journal of Rock Mechanics and Mining Sciences and Geomechanics Abstracts 31(1): 23-34.

Simpson, R.W., 1997. Quantifying Anderson´s fault types. Journal of Geophysical Research 102 (B8): 17,909-17,919.

Sissingh, W., 1998. Comparative Tertiary stratigraphy of the Rhine Graben, Bresse Graben and Molasse Basin: correlation of Alpine Foreland events. Tectonophysics 300: 249-284.

Sittler, C. and Sonne, V., 1971. Vorkommen und Verbreitung eozäner Ablagerungen im nördlichen Mainzer Becken. Neues Jahrbuch für Geologie und Paläontologie, Mitteilungen: 372-384.

Sittler, C., 1965. Le Paléocène des fossés rhénan et rhodanien. Études sédimentologiques et paléoclimatiques. Mémoire du Service de la carte géologique d'Alsace et de Lorraine 24, 392 pp.

Spang, J.H., 1972. Numerical method for dynamic analysis of calcite twin lamellae. Geological Society of America Bulletin 83 (1): 467-472.

Sperner, B. and Ratschenbacher, L., 1994. A Turbo-Pascal program package for graphical presentation and stress analysis of calcite deformation. Zeitschrift der deutschen geologischen Gesellschaft 145: 414-423.

Stäblein, G., 1968. Reliefgenerationen der Vorderpfalz - Geomorphologische Untersuchungen im Oberrheingraben zwischen Rhein und Pfälzer Wald. Würzburger geogr. Arbeiten, 23: 1-183.

Stapf, K.R.G., 1988. Zur Tektonik des westlichen Rheingrabens zwischen Nierstein am Rhein und Wissembourg (Elsaß). Jahresberichte und Mitteilungen des oberrheinischen geologischen Vereins, N.F. 70: 399-410.

Stein, R. S., Barka, A. A. and Dieterich, J. H., 1997. Progressive failure on the North Anatolian fault since 1939 by earthquake stress triggering. Geophysical Journal International 128: 594-604.

Stein, S. and Wysession, M., 2003. An introduction to seismology, earthquakes, and earth structure. Blackwell Publishing, 512 pp.

Straub, E.W., 1962. Die Erdöl- und Erdlagerstätten in Hessen und Rheinhessen. Abhandlungen geologisches Landesamt Baden-Württemberg 4: 123-136.

Streit, J.E. and Hillis, R.R., 2004. Estimating fault stability and sustainable fluid pressures for underground storage of CO_2 in porous rock. Energy 29: 1445-1456.

Tenzer, H., Mastin, L. and Heinemann, B., 1991. Determination of planar discontinuities and borehole geometry in the crystalline rock of borehole GPK-1 at Soultz-sous-Forêts. Geothermal Science and Technology 3 (1-4): 31-67.

Tenzer, H., Budeus, P. and Schellschmidt, R., 1992. Fracture analyses in hot dry rock drillholes at Soultz and Urach by Borehole televiewer measurements. Geothermal Resources Council Transactions, Davis, CA, pp. 317-321.

Tenzer, H., Schanz, U., Homeier, G., 2001. Main Results and further research work at HDR-Testsite of Urach Spa Development of a HDR Demonstration Pilot-Plant. International Summer School on Direct Application of Geothermal Energy, Conference Proceedings, Chapter 3.2, 5 pp.

Terzaghi, K., 1923. Die Berechnung der Durchlässigkeitsziffer des Tones aus dem Verlauf der hydrodynamischen Spannungserscheinungen. Akademie der Wissenschaften in Wien, Sitzungsberichte mathematisch-naturwissenschaftliche Klasse, part IIa 132 (3/4), pp 125-138.

Tesauro, M., Hollenstein, C., Egli, R., Geiger, A., Kahle, H.-G., 2005. Continuous GPS and broad-scale deformation across the Rhine Graben and the Alps. International Journal of Earth Sciences (Geologische Rundschau) 94 (4): 525-537.

Tesauro, M., Kaban, M. K., Cloetingh, S.A.P.L., Hardebol, N. J., Beekman, F., 2007. 3D strength and gravity anomalies of the European lithosphere. Earth and Planetary Science Letters 263: 56-73.

Théobald, N., 1934: Les alluvions du Pliocène supérieur de la région du Sundgau. Bulletin de la Société Industrielle de Mulhouse 101: 1-36.

Tietze, R., Neeb, I., Walgenwitz, F. and Maget, P., 1979. Geothermische Synthese des Oberrheingrabens (Bestandsaufnahme) - Synthèse géothermique du fossé rhénan supérieur (Etat des connaissances). Freiburg, Strasbourg, Geologisches Landesamt Baden-Württemberg, Service Géologique Régional Alsace, 51 pp.

Turcotte, D.L. and Schubert, G., 2002. Geodynamics. Cambridge University Press, Cambridge, 347 pp.

Twiss, R.J. and Moores, E.M., 1992. Structural Geology. Freeman Co., New York, 532 pp.

Valley, B. and Evans, K.F., 2007. Stress state at Soultz-sous-Forêts to 5 km depth from wellbore failure and hydraulic observations, Thirty-Second Workshop on Geothermal Reservoir Engineering, Stanford University, Stanford, California.

Van Balen, R.T., Houtgast, R.F., Van der Wateren, F.M., Vandenberghe, J. and Bogaart, P.W., 2000. Sediment budget and tectonic evolution of the Meuse catchment in the Ardennes and the Roer Valley Rift System. Global and Planetary Change 27: 113-129.

Vigny, C., Chery, J., Duquesnoy, T., Jouanne, F., Ammann, J., Anzidei, M., Avouac, J.-P., Barlier, F., Bayer, R., Briole, P., Calais, E., Cotton, F., Duquenne, F., Feigl, K.L., Ferhat, G., Flouzat, M., Gamond, J.-F., Geiger, A., Harmel, A., Kasser, M., Laplanche, M., Le Pape, M., Martinod, J., Menard, G., Mayer, B., Ruegg, J.-C., Scheubel, J.-M., Scotti, O. and Vidal, G., 2002. GPS network monitors the Western Alps' deformation over a five-year period: 1993-1998. Journal of Geodesy 76: 63-76.

Villemin, T. and Bergerat F., 1985. Bruchtektonik und känozoische Paläostressfelder am NE-Rand des Oberrheingrabens. Oberrheinische geologische Abhandlungen 34: 63-87.

Villinger, E., 1998. Zur Flußgeschichte von Rhein und Donau in Südwestdeutschland. Jahresberichte und Mitteilungen der oberrheinischen geologischen Vereinigung, N.F. 80: 361-398.

Walter, R., 1995. Geologie von Mitteleuropa. Schweizerbart, Stuttgart, 566 pp.

Wells, D.L. and Coppersmith, K.J., 1994. New Empirical Relationships among Magnitude, Rupture Length, Rupture Width, Rupture Area, and Surface Displacement. Bulletin of the Seismological Society of America 84(4): 974-1002.

Wenzel, F., Brun, J.P. and ECORS-DEKORP team, 1991. A deep reflection seismic line across the Northern Rhine Graben. Earth and Planetary Science Letters 104: 140-150.

Wirsing, G., Luz, A., Engesser, W., Koch, A., Elsass, P. and Perrin, J., 2007. Hochauflösende Reflektionsseismic auf dem Rhein und Rheinseitenkanal, LGRB-Fachbericht 01/07, Freiburg i. Br.

Wittmann, O., 1955. Bohnerz und präeozäne Landoberfläche im Markgräflerland. Jahresberichte des geologischen Landesamt Baden-Württemberg 1: 267-299.

Worum, G., van Wees, J.D., Bada, G., van Balen, R.T., Cloetingh, S.A.P.L. and Pagnier, H., 2004. Slip tendency analysis as a tool to constrain fault reactivation: A numerical approach applied to three-dimensional fault models in the Roer Valley Rift System (southeast Netherlands). Journal of Geophysical Research, 109 (B02401): 10.1029-10.1044.

Zeis, S., Gajewski, D. and Prodehl, C., 1990. Crustal structure of southern Germany from seismic refraction data. Tectonophysics 176: 59-86.

Ziegler, P.A., 1990. Geological Atlas of Western and Central Europe. 2nd Ed. Shell International Petroleum, Maatschappij, distributed by Geological Society, London, Publication House, Bath, 239 pp.

Ziegler, P.A., 1992. European Cenozoic rift system. Tectonophysics 208: 91-111.

Ziegler, P.A., 1994. Cenozoic rift system of western and central Europe: An overview. Geologie en Mijnbouw 73: 99-127.

Ziegler, P.A., van Wees, J.-D., Cloetingh, S., 1998 Mechanical controls on collision-related compressional intraplate deformation. Tectonophysics 300: 103-129.

Ziegler, P.A., Schumacher, M., Dèzes, P., van Wees, J.D. and Cloetingh Sierd, A.P.L., 2004. Post-Variscan evolution of the lithosphere in the Rhine Graben area: constraints from subsidence modelling. In: M. Wilson et al. (Editors), Permo-Carboniferous Magmatism and Rifting in Europe. Geological Society Special Publications, London, 223: 289-317.

Ziegler, P. and Dèzes, P., 2005. Evolution of the lithosphere in the area of the Rhine Rift System. International Journal of Earth Sciences, 94(4): 594-614.

Ziegler, P.A. and Dèzes, P., 2006. Crustal Evolution of Western and Central Europe. In: Gee, D.G. & Stephenson, R.A. (eds) European Lithosphere Dynamics. Geological Society, London, Memoirs, 32, 43-56.

Ziegler, P.A. and Dèzes, P., 2007. Cenozoic uplift of Variscan Massifs in the Alpine foreland: Timing and controlling mechanisms. Global and Planetary Change 58, (1-4): 237-269.

Zoback, M. D., 2007. Reservoir Geomechanics. Cambridge University Press, pp 464.

Zoback, M.D. and Healy, J.H., 1984. Friction, faulting, and insitu stress. Annales Geophysicae 2: 689-698.

Zoback, M. D., Zoback, M. L., Mount, V. S., Suppe, J., Eaton, J. P., Healy, J. H., Oppenheimer, D., Reasenberg, P., Jones, L., Raleight, C. B., Wong, I. G., Scotti, O. and Wentworth, C., 1987. New evidence on the state of stress of the San Andreas fault system. Science 238: 1105-1111.

Zoback, M.D. and Healy, J.H., 1992. In situ stress measurements to 3.5 km depth in the Cajon Pass scientific research borehole: implications for the mechanics of crustal faulting. Journal Geophysical Research 97: 5039-5057.

Zienkiewicz, O.C. and Taylor, R.L., 1994. The finite element method, Vol. 1&2, McGraw-Hill Book Company, London. 752 pp.

Zienkiewicz, O.C., Taylor, R.L. and Zhu, J.Z., 2005. The Finite Element Method. Its Basis and Fundamentals. Butterworth-Heinemann, 752 pp.